The Cosmic Microbe

Towards a Becoming Life on Planet Earth

Heather Couzyn

Correspondence concerning *The Cosmic Microbe* should be addressed to:

Mrs.H.E.Couzyn
99 Ninth Street
Parkhurst
JOHANNESBURG 2193
South Africa
Telephone: (011) 442-7688

ISBN 0 9513045 9 3

Published in 1995 by
Headway Books, Birch Hagg House, Low Mill, York, YO6 6XJ

Cover Design: Siobhan Brownlow.

Printed in Great Britain by Redwood Books, Trowbridge, Wiltshire.

DEDICATION

For planet Earth, for people, and for all living things.

CONTENTS

(Key concepts are marked with an *)

PART SIX

TOWARDS SYNERGETIC COHERENCE IN HUMAN ACTIVITY

(A Look at Current Thinking)

PART SEVEN

GAIA AND THE MICRO-CHIP

PART EIGHT

MICROCHIP ECONOMICS IN ACTION

Book III

PART NINE
STEPS TOWARDS TRANSITION

PART TEN
THE INFORMATION SOCIETY

CODA

FOREWORD

This is a book of exhilarating hope and insight.

It presents, with convincing logic, a deep analysis of the human condition, its probably origin, its profound complexity and, in keeping with the emerging technology of the information age, a positive future trend.

Wide research and clear reasoning reveal the good that can result from our growing knowledge of nature's behaviour, intricacy and beauty, and the extent to which electronic techniques can be used to implement this good in society.

Against this background the author presents pointers for the future which the critical reader will find both reasonable and reassuring, at the same time taking him on a voyage of discovery which rivals a good detective story.

He is led from the known world of the physical sciences to the logical conclusion of social responsibility, mutually advantageous ethical behaviour and the final emergence of a friendly society drawn together by the microchip.

This author's contagious conviction that the human condition must and will improve makes for an inspiring work of art and science that should be read and studied. From the fusion of economics with the information sciences there emerges a workable concept in tune with Lovelock's original Gaian hypothesis and the author's own insistence that a benign society is possible.

From having spent many years working on solid state devices as forerunners of the microchip, and two decades with the emergent power of the very-large-scale integrated circuits, I believe the author's hope, (based on the benign influence of these powerful systems), although optimistic, is not unrealistic.

It is true that the complexity and computational speed of the newer generations of logic arrays are expanding at breathaking rates, but the microdimesions now under test are severely straining the system-reliability attainable. Also the architecture of future information-handling systems may differ appreciably from the predicitve architecture currently still in use.

Attempts at probable computational arrays, and fuzzy logic applications still fall far short of even the simplest intuitive deduction of which a living brain is capable. But this does not rule out their staggering possibilities for the future.

There seem to be other challenges in the realization of systems with storage density approaching the living brain, not the least being thermal constraints.

In the present mode of information storage by transfer of electrical charge between interconnected nodes, quantum mechanical considerations point to a severe problem related to the access and retrieval-time of discrete quanta of information. Consideration of the density of information belived to be stored in the DNA-strucure, and the apparently long time required to 'programme' this memory, does tend to underscore the possible existence of hurdles still to be scaled.

The apparently minute energy levels involved in the electrical activity of thought processes, and our inability to monitor them without introducing massive disturbances, will probably stress our recording techniques to their very limits. However, since our knowledge of, and thence our models of, information storage are at best still relatively crude, with further insight present obstacles may disappear.

The simple fact that the human brain in all its wondrous complexity consumes energy at a rate of only about 20 watts, is a clear pointer to our still limited grasp of the nature of living cell information-handling.

It is important that we do not only try what is known as 'logical' reasoning to explain what we observe. Innovation may tap a higher level of intelligence, as already hinted at by the 'emergent properties' quoted by the author.

The beauty and promise of the model presented here tend to strengthen the notion that information engineering, in its widest context, is a goal eclipsing, in its potential impact on mankind, even the present wonders of space exploration.

The Cosmic Microbe is exceptioanlly well researched. It contains enormous amounts of condensed information and will reward the reader with a song in the heart and soundly based hope for the future.

L van Biljon D. Sc. Eng.

AUTHOR'S PREFACE

> *'We count because each one of us - ordinary men and women carry the permanent God within us and are therefore capable of moving the whole species, the whole enterprise forward from where it was when we were born.'*
>
> Anna Brito

I hope this book will strike a chord in my fellows, and they will forgive its weaknesses. I think it says things which need to be said. I think we all care. And if some of it proves difficult, it is because I have found it so, exploring realms I knew too little about, yet engrossed and absorbed; too much in love with the Earth to let it go. Besides, it is fun in there.

Few of us are happy today when we look at our world. Few can fail to question where we've gone wrong. The prophets of doom are proving so tragically right. The crisis they've been predicting for years is at our door, and it is worse than they knew.

Most of us are aware of these things. Our foolishness could spell the demise of mankind, and each of us feels like an ant in the path of a train.

We can shut our eyes to this knowledge, or we can respond, but it won't go away. My book, *The Cosmic Microbe*, is what it engendered in me. Grain of sand though it be, I hope it will help. It is the only response I know how to make.

Many other authors are doing the same; many with far more knowledge and skill than mine, but surely with no more concern. Our common cause is the future of man and of life; our common love this beautiful fragile Earth. So I offer my book in the stirring knowledge that each of us counts. The future, if we are lucky enough to have a future, will rise like a phoenix from the caring and action of each individual man.

Book I

The evolution of my book has been a long one. It started more than forty years ago in a search for understanding that led me down many and diverse paths, starting with literature and the arts, moving on through physics to biology; dallying in philosophy; passing unwillingly to politics; ending in the social and economic world. I still have no more than a layman's superficial acquaintance with any of these subjects. But because I was born with a faculty for recognizing relationships: their apparently disparate scraps of knowledge tended throughout my life to present themselves in one coherent whole. Each new thing I learned would fuse immediately with everything else, engendering in the process a host of unanswered questions; further exciting puzzles were there to explore. The need to understand had become a way of life. The world had become one world.

This capacity to integrate an infinity of discrete elements is at once my strength and my weakness: my strength because that is how the Earth, and indeed the universe, appears to be structured; but also my weakness in communicating with my fellows, most of whom seem not to look at life in quite the same way. Their issues are clearer, clean-cut, uncomplicated, and this, for me, often makes them untrue. But they are much easier to express than the shimmering webs of thought that hold me in thrall. Mine are more fun! But for the precision of specialist knowledge, I have a respect that sometimes amounts to awe.

The relationships of my own cerebral world are so interlocked I still search in vain for an appropriate word for them. Neither 'networks' nor 'webs' will do. Both are linear in essence, and over-used. And neither conveys quite the dynamic of interacting energies by which everything happening anywhere in the universe is reflected in everything else. Down here on Earth this includes our feelings and thoughts - abstract relationships as well as functional and spatial ones, and Earth can only be rightly conceived to include them all. Our thoughts and our feelings are quite as real as our flesh, and Earth's totality embraces them. It all adds up to a planet as live as ourselves; a living creature subject to natural law; an Earth that is able to suffer and able to die; complex and mysterious as we; majestic and divine.

This notion will take some swallowing at first. I know it is rather a large statement, and one which will take sound argument to justify. But for me it is truth. I hope to show that because of it, our plight is not so hopeless as it appears.

Like all concerned people today, I know that pursuit of wealth has all but destroyed the planet on which we live, and on whose ability to sustain itself our survival depends. I also know that while informed people all over the world are striving to reverse this trend, they find themselves shackled at every point by economic constraints. Is there any way out? For a long time the quest seemed hopeless and even ludicrous. But against this rather disheartening background I plodded on. Increasingly, as I studied and read, economic restraints began to lose their relevance. In a world equipped to operate as a living thing, all our parameters had changed. Information could take the place of wealth. The tools were there to our hand. Money, I realized at last, had outlived its usefulness. We now had a wider vision, and far far better tools. Bringing them together would release a new dynamic in human affairs.

And so conviction grew that the laws governing our frightening world were not immutable. Our problems were relatively recent - and man-made, ranging from overpopulation and pollution to ignorance, stupidity and greed. As such, they could be rectified by man. They concerned identifiable causative factors of power and capital which had brought us as far as they could usefully go. What we now needed was a shared understanding of practical alternatives, a conviction that each man's individual contribution counts, a determined exploration of possible routes, and profound care for planet Earth. So much for Book I.

Book II

Book II is based on the perception that the key to useful possible routes lies with computers. Living systems operate on a closely integrated network of information by which their activities are constantly modified and renewed through a process of feedback from all their parts. The development in the last few decades, at exactly the right historical moment, of humanity's latest toy, the computer, will give mankind the power to do the same. Computers are crucial to this hope. Their significance is this:

If Earth is a sensate creature like ourselves, we can, with the aid of computers, begin to organize society in conformity with known cybernetic principles. Fed into a network of computers, and thereby put in touch with each other, the input from each one of us can begin to have a direct and immediate impact on the thoughts and lives of all of us. Using the process of feedback common to living systems, we can begin to get organized. We can sort out our social memes and get to know each other; find out what each man really wants and loves and put it his way. In particular we can develop a free, self-regulated, need-orientated, information-based economy in which each man's self-interest and initiative can be fulfilled, without recourse to value systems, physical means of exchange, or capital. So the question Book II attempts to engage is this:

How, in the light of a living, sensate Earth, can the socio-economic structure of society be modified or recast to give everyone at least a sporting chance, and ennoble each man's life?

Answers are found in a marriage between the known biological principles governing living organisms and the immense integrating power of the computer.

In this convergence of two powerful principles, money gives way to information as the driving force of human activity, and society moves towards a more truly human organization of its affairs.

Book III

Book III explores some of the transitional steps toward such reorganization, and attempts to show that current trends in both business and politics are already moving towards them. Indeed it shows that business itself will be a prime mover in the healthy transformation of society, and that the present chaotic state of world politics is a necessary prelude to the new order.

Earth, if she survive at all, will be a good Earth.

Heather Couzyn
October 1994

AUTHOR'S NOTE

The word 'man' is used throughout this book in its widest sense of a human being, and of mankind in general.

It is used in this way purely as a matter of convenience, and has no sexual connotations of any kind.

Without exception it refers to every man, woman and child on Earth. Feminists please take note.

Book I

Gaia, the Earth, is a living cell

PART ONE

Selfhood and the Universe

CHAPTER ONE

A Personal Philosophy

In my understanding the planet Earth is alive, like you or me. Not only is it alive, it is moving fast toward self-consciousness. It is a cell, an egg. It is an embryo. It is a cosmic creature, burgeoning into life - closer than breathing, nearer than hands and feet.[1] We are flesh of its flesh: cells of its marvellous multiple body; neurons in its evolving brain; core of its mind.

One cannot know this truth with the mind alone. It is not an assessment of fact, that computers could emulate. It is an instinctive thing, race-memory, going right back to the roots of the human line, to the primal cells and beyond. Earth remembering from whence she came. Earth in a cosmic system rooted in God. This is my starting point.

I care immoderately about this. I am part of the force that somehow gave it birth. So caring is innate in the cosmos. I care because I carry its patterning. I am an elemental speck of an elemental speck in the unimaginable immensity, but still life of its life, substance of its substance. Whatever *was* therefore, before the beginning, is somehow built in to my genetic and molecular memory, and into every particle and cell and creature on Earth, whether we recall it or not. In a very real sense, we were there.

Every one of my cells reflects this totality. Every atom and planet and star carries somehow the stamp of its beginnings. If we could learn to read ourselves, we'd know whence we came. We'd know where we fitted in the universe, and what we were. The way things are we have to trust to our intuition, and guess, using the thoughts that rise most persistently to our minds and feel most true. It goes right back to time before Time. Before creation. It goes back to God. But the thing that surfaces from this wellspring of being for each of us is unique to himself. According to what he is, he will comprehend.

I've felt this unity all my remembered life. I moved toward it blindly, like groping in fog, feeling with my spirit for its inwardness, but lacking the data that could link it to my mind. My mind said I was here, and the world out there. But in my inner space they were only one. So I lived two lives. Perhaps almost everyone does: one fragmented, that people know him by; and one inside that is whole, where the images form. The inward one flames incandescent at times, as piece after piece of the pattern falls into place, and mind and spirit are one.

For one splendid year when I was six years old, we lived on a farm. It was a happy farm. There were cows and horses and pigs, and a kingly red-gold rooster with flaming wings. The creatures were loved. And there were places where a child could go to dream.

We used to climb a tree to the dairy roof, my sister and I, a secret hide-out, out of sight of the house, where time passed by unnoticed. The sounds of the farm would filter through to consciousness. You could hear the clanking of buckets at milking time and the cows lowing, calling to their calves. Sunlight warmed our shoulders and brown young arms, spilled on down to the roof, and slanted in dusty shafts of light to the haystack below. We children had 'helped' to build that stack, riding high on the wagons at harvest time, and wielding pitchforks dangerously.

From where we sat on the roof the hay was a tempting invitation to jump, if you were brave; a grassy landing halfway down to the ground. We had considered this. Or rather my sister had. For me it was remote enough to stay on the edge of things. It did not intrude on my world. Fragrance of cut hay was drifting up. The late sun filled the dancing air with gnats. I was the sun, the sky, the earth, the down-soft clouds. Personal boundaries blurred, seamless as time. Life was pure experience, not personness.

'Jump!' my sister commanded suddenly, and jumped.

I jerked back inside my personal skin, bemused and afraid. Twelve feet up was suddenly very high, and I just a frightened child. I liked it better where I had been before.

All my life I have had this trouble with self, my edges dissolving into the universe, consciousness drifting about like mist on the lake and never still. I have never known who I was, and I didn't care. Just being it was the absorbing, wonderful thing.

Was I my thoughts or my person? My feelings or those of the people I loved? They all felt equally real to me. I seemed to respond to them like the strings of a harp. Was the life I sensed in a stone really there, or in my mind? To know, I had only to take it in my hand.

I used to try very hard to isolate my 'self'. I would tell myself over and over: "I exist. I am conscious, contemplative. I am myself. I matter." It didn't work. I am as much the 'other' most of the time as I am I. Transient. Unfenced. Selfhood insufficiently self. The universe flows through me like the solar wind. In an age when we're more concerned with selfhood than ever before, this makes me a maverick. Personality matters. Character matters. Identity matters. Selfhood is the very key to life. I know this with my mind, but I seldom feel it. When I do, there is pain. I am more comfortable with boundaries blurred; intense and interested and wonder-filled; part of a larger self which is the universe. More closely still, I am the Earth, pulsing into its wider consciousness; consciously flowing; consciously fulfilled. I am happy. I like it this way. I think it is a way of things to come. Selfhood and consciousness. Two wings flying.

I am not alone in this. Many creatures in nature work this way, even the primitive ones. Most of all, perhaps, the primitive ones. Typical is a strange little forest slime, an 'amoeba' when living it's separate 'self'. It grazes away on the forest floor, apparently unaware of its destiny; discrete, self-contained, accountable to none. At moments like these it is as fully a self as we.

Then food gets scarce. You'd expect a frantic scramble to get there first, survive at all costs, each man to himself. But not at all. The picture changes dramatically. Perhaps half a million of these so-primitive cells converge together spontaneously, directed, it seems, as though by a single mind, and join together, forming an animal. And as though this were not enough, each cell adjusts to the new, now corporate self, and fills a specialized role in the new creation. All are now no more than the flesh of the slug, a forest slime.[2]

The slug moves about for several weeks in search of new feeding grounds. It has a front and a rear end, negotiates obstacles, forms a fruiting body to reproduce itself, and has some kind of consciousness. Imagination balks at how this must feel, but this is not the end of the story.

When the fruiting body sheds its spores, each can split open to liberate a tiny new creation - not, mark you, a slug this time, but a single, self-sufficient, autonomous amoeba that recommences the life of a single cell as though it had never heard of a corporate life.

This plays havoc with identity. But is this what a seamless universe really means? Do I, at a mental level, perform the same feat? And which of my separate states is really me? I like the seamless one best. It is a lovely, lovely way to live. Nothing I feel as a person can touch the way I feel when the seams dissolve and I am open to life. The rhythms of life transport me and lift me up. I find myself worshipping. Merged in the mind of a living, sensate Earth, my selfhood becomes a transcendental thing, aware with a sense more potent than its own.

For many people, no doubt, the thought of sharing like this in a corporate self would be anathema. It isn't like that for me. I am built, like the slug, of separate cells, and am aware of them. For the span of my natural life they are this composite 'I', my body and mind, comprising their billions of interrelated lives. We breathe and think and love together, a new kind of creature, somehow still themselves. My cerebral powers are awesome compared with theirs. My corporate self transcends their togetherness. If Earth should bring to fruition a corporate mind, I'd share in this wonderment: resound perhaps like a Bach to her symphonies, see like a Rembrandt or an Angelo; feel the compassion of a son of God.

May it be soon.

Is it enough to be part of this splendid thing? to share in the Earth's majestic coherence of mind? Rambunctious, vital, beautiful, she is an embryo burgeoning into life. Transition is hard and painful, but here we are. Will Earth burst through to life in spite of us, or do we have to take a hand in it?

I try to laugh the thing off, but it doesn't work. Ridiculous microdot in the order of things, I am flesh of her flesh, caught in the turmoil and anguish of her transition. Can I contribute by my conscious will?

I'm more than a little embarrassed by how I feel. It does not do to be so passionate. But if there is truth in the way I see my world, we need to engage ourselves. The thinking and will of people who really care will determine the quality of global mind.

I want to be germinal to this. I want to be part of the growing body of people who feel the same way: people who've loved; people whose books I have read, or whom I only sense out there beyond the circle of my known friends: people whose total caring for planet Earth has become the prime motivation of their lives. I am trying to respond, I think, to nature's rhythms; to beat with the overall pulse of her awakening self; to generate, if I can, a noble resonance in the dynamo.

Each man lives perforce in his own cerebral world, self-created, unique to himself alone. He can expand or contract it, enrich or impoverish it, sustain or destroy it. It houses his mind and spirit through his mysterious brain. His web of cerebral tissue links him to the Earth and to the stars. It vibrates to his personal frequency. It is his access to the universe. If he search long enough, and hard, he will find its centre, his personal dynamo, the core of conviction and vision that fires his soul. This is mine: I foresee an Earth which is known for its love as a salmon is known to swim or eagle soar. There is no pain can negate its validity. To the furthest reach of my powers I will nourish it. I am committed to this. My will is focused on this single thing. And if it is more than my potential permits, I will transcend myself, and so die happy, trying.

PART TWO

The Evidence for a Living Earth

CHAPTER TWO

2.1 Pointers in Physics

Summary: Everything in the Universe is interconnected.

Is there a basis of fact for my wild ideas? There does seem to be so at last, as science draws ever closer to mysticism, and many new facets of truth begin to converge. But mine are a layman's views, not a scientist's. The source-books of science were way above my head. They drew me as a moth is drawn to flame, so I read them anyway, and sometimes recognition flared, burning into the tinder of my mind with tongues of fire. Pieces would suddenly flame together, illumined; incandescent molecules of thought. And as they flamed and fused,excitement soared.

Conclusions were always intuitive. I did not usually know, as I read, why certain things rushed out from the page and blew my mind. I only knew they were important to me. A phrase would simply *feel* right, or pass me by, however logical its argument. Then I would say 'Thanks, friend,'and mop it up, and gratefully digest its nourishment. I grasped at it as a starving man grasps food, impelled by an urge I could not explain nor control. I only knew that it fed the hunger in me. Someday, somehow, in the network of circuits forming in my brain, it would find its usefulness. I clung to this. Today in computer terms we'd call it 'bottom-up'.

One day I stumbled on this:

> The inertial properties of terrestrial matter are determined by the total mass of the universe around us.[1]

My senses reeled. The total mass of the universe, stars, planets, quasars, red giants - the lot! All working together in concert to create that subtle inertial balance that I must shift when I try to push my motor car. I, when the wheels start rolling, engage the universe.

It didn't seem possible at first, but it brought my heart pounding into my throat. Here was the data I'd sought for through all those years. Nobody seemed to know how it worked, but what it clearly implied was that everything anywhere, influenced *everything else, all at once,* throughout the universe. Nothing could now be conceived in separateness. Everything acted, and was acted upon, not just by adjacent entities, but by the inter-resolving mass of the cosmos itself. All must now be seen in relationship. The energy patterns of the universe are encoded in every part of it.

David Bohm speaks of 'implicate' and 'explicate' order,[2] but actually says the same thing from a different angle.

Explicate order is the universe we see around us. Implicate order the unseen and physically immeasurable.

> At the level of implicate order, every part of the universe contains the whole universe enfolded within it.[3]

If you think of matter as composed of particles, this is a daunting concept to visualize. It is easier when you think of energy fields, with waves pulsing in every direction at once, cancelling each other out, reinforcing each other like waves in a pool, each affected by all. At a point of high wave density, matter is manifest. The subatomic particles of matter are thus like moments in the infinite patterning of energy waves. Where they peak, implicate gives way to explicate, and matter appears.

Other people's philosophy is usually rather boring, but don't go away. Wholeness is crucial to a living Earth, and after a notion as bizarre as that, anything can happen!

It is all a little easier when you think of the hologram, a three-dimensional image produced by laser light. Lasers have a familiar ring today, whether we understand them or not, so they're not so daunting.

'In a normal photograph,' says Peter Russell,[4] 'each point in the photograph is a particular part of the final image, and for the image to be seen correctly, all the points need to be in the correct position.' (So far, so good!) 'In a hologram, on the other hand, each point on the photographic plate records data about the whole image. Every part of the image is encoded in every part of the plate.' (This sounds improbable, doesn't it? But it's true!) And, says Russell, since any region of the plate contains information about the whole image, it can give rise to the *whole* image...' Substitute 'universe' for 'image' and you have a graphic visual picture of the behaviour of light in space. In Lyall Watson's unforgettable words:

> Any old point in space will do. They all embrace everything happening everywhere.[5]

Watson suggests that theoretically one could produce holograms including microscopes and telescopes that would give you a view of everything from single cells to double stars. 'Then you could carry one small chip around as a charm, a sort of concentrated cosmos imprisoned in a signet ring.' I'd like that! But he goes on to say that all that effort would be pointless, because in essence that is what we already have in every cell of our bodies. 'That is what life means. It is the lens in the hologram, our connection to the cosmos. Direct dialing to anywhere in the universe.'[6]

This is so exactly how I feel, I'd like to shake him by the hand! It is only possible because, at the subatomic level, our concepts of space, time, matter and logic no longer apply. The subatomic particles of the physicist's bubble chamber behave in a way totally irreconcilable with common sense. It all reads very much like a fairy tale. Electrons jump from one orbit to another in the most delightful way without passing through the space between them. Two particles may collide in the bubble-chamber, and break into pieces, and these pieces may turn out to be bigger than the original particles! Or a photon, the elementary unit of light, which has no mass, can give birth to an electron-positron pair which does have mass. A diagram tracing a bubble-chamber photograph of subatomic events holds more magic than Kandinsky or Klee. It is sheer fantasy.

I can take just so much of this kind of thing before I drown. But the physicists hadn't finished with me yet. *Bell's Theorem*[7] was still to come. It came at me out of the blue like a lightning blast. *Bell's Theorem* states that when two particles have interacted and then flown off in opposite directions, interference with one particle will instantly affect the other particle, *regardless of the distance between them*. The implications of this astounding revelation surpass imagination. Once again they suggest a fluid, unbroken dynamic of energies incapable of analysis into separately and independently existent parts. And this dynamic includes you and me - every atom of our bodies, every particle of every breath we breathe, every electro-magnetic impulse of our nervous system, every thought we think. Everything is shaped, quite literally, by *everything*. Every moment recreates the universe.

Wild thoughts these, jumping whole realms of argument, transposing words from a physicist's mind into things that he has not said. Calm down. Come back to earth. Be practical. What are the implications for us *here*, in our daily lives, and why should a concept so dimly perceived excite me so? I tried to come back from dreaming and work it out. Either everything mattered, or nothing. I wasn't sure which. But unless these things could show us how we could live more sensibly, they were only for physicists. Intellectual excitement alone was not enough. There were *people* down here.

Start at the other end then. If creation was wholly enmeshed like this, and its laws applied in the furthest reaches of space as they did here on earth, there must be a unity of living things as well - people and creatures and plants and trees co-existent with universe. The seamlessness I had sensed must really be true, and life itself must be a unity. Since we were alive, then so must the universe be, and so must the Earth. It made me feel less lonely and foolish at last, and less of a changeling. This knowledge could change the pattern of our lives.

CHAPTER THREE

2.2 Living Systems

Summary: Planet Earth is recognizable as a living system by the many mutually sensitive interdependent parts of which it is composed, and by a number of physiological attributes common to such systems.

For the planet Earth to be physically recognizable as a living entity, it would need to be composed of mutually sensitive interdependentparts. It should be able to maintain for itself a relatively constant temperature, humidity, pressure, chemical equilibrium and metabolism, all of which criteria it does indeed appear to fulfil. There's been a great deal of research in this field in recent years. One of the leading researchers was Dr James Lovelock, British chemist and inventor, who, in the early 1960's, served as consultant to a team at the California Institute of Technology working on methods of life-detection on Mars. Lovelock realized that life-forms on alien planets might have evolved from a different set of chemical constituents from our own, and might therefore not be recognizable by tests modelled on our own chemical system. Pondering this difficulty it occurred to him that since one of the hall-marks of living systems was their apparent contradiction of the universal law of entropy, it might be possible to recognize the presence of life in an alien environment by searching for signs of entropy-reduction.

According to the *Second Law of Thermodynamics*, all energy will eventually dissipate into heat which will be universally distributed and will no longer be available for the performance of useful work. Chemists measure the rate of this dissipation of a system's energy as a function of 'entropy', which can be precisely expressed in mathematical terms. According to the *Second Law of Thermodynamics*, 'whenever a reaction occurs, the usable energy in the reaction system must decline. In other words, the level of energy within a system tends to run downhill.'[1] Universally applied, this implies the predestined and inevitable rundown and death of the universe.

Living systems, however, show no signs of running down. On the contrary, they succeed in maintaining a high degree of internal order. The whole history of evolution of living forms is one of increasingly complex order in the face of an apparently universal trend towards disorder.

Assuming that life on any planet would be bound to use fluid media - oceans, atmosphere or both - as conveyor belts for raw materials and waste products, it occurred to Lovelock that some of the activity associated with entropy reduction within a living system might spill over into the conveyor belt regions and alter their composition. The atmosphere of a life-bearing planet would thus become recognizably different from that of a dead planet. Using our own planet as a model, Lovelock and his colleagues embarked on a widespread atmospheric analysis. Their results convinced them that 'the only feasible explanation of the Earth's highly improbable atmosphere was that it was being manipulated on a day-to-day basis from the surface, and that the manipulator was life itself.'[2]

Thus Lovelock's brilliant perception of entropy reduction as a life-detection mechanism was to open the way to recognition of the Earth itself as a living system.

Later work for Shell Research Limited on the possible global consequences of air pollution gave Lovelock the opportunity to carry this hypothesis a step further. The atmosphere of the Earth behaved so contrary to the recognized laws of chemistry that it looked like 'a contrivance put together co-operatively by the totality of living systems to carry out certain necessary control functions.'[3]

All the living organisms on Earth, together with the air, sea, and land network, seemed to Lovelock to compose one enormous integrated system that somehow kept itself aware of the conditions vital to survival and took highly complex measures to ensure their continuance. Lovelock called this system Gaia, in honour of the ancient Greek Earth goddess Gaia (or Ge). His results, published in 1979, present Gaia as 'the largest living creature on Earth,'[4] and stop short of including the planet itself as a living organism.

To me it seems that the cyclical nature of biological processes precludes such a distinction. We are *continuous* with earth and sky, inseparable from them in our living as in our dying, just as we are inseparable from each other. We breathe the same air, drink the same rain; inhale oxygen exhaled by plants, which in turn use the carbon dioxide exhaled by us. We are nourished by chemicals out of the soil; and return through our detritus and our dying whence we came, as does every living creature and plant on planet Earth. 'We did not come into this world,' says Lyall Watson. 'We came out of it, like buds out of branches and butterflies out of cocoons. We are a natural product of this Earth, and if we turn out to be intelligent beings, then it can only be because we are fruits of an intelligent earth, which is nourished in its turn by an intelligent system of energy.[5]

Clearly then, for Watson as for me, if Gaia turns out to be alive in the sense that Lovelock suggests, she must embrace the whole skeletal framework, as well as all the living 'cells' of the biosphere. The planet in its entirety must be a living thing.

CHAPTER FOUR

2.3 Dissipative Structures

Summary: Open chemical and biological systems in continual interaction with their environment (known as Dissipative Structures), are found to have the capacity, under certain conditions, to generate a new dynamic order from disorder.

The Earth is such a system.

It would be difficult to conceive of anything less indicative of global interrelatedness than human behaviour. If Gaia does indeed have the regulatory powers that Lovelock suggests, why do we lumber about maiming and killing each other, prostituting the Earth, jeopardizing our very survival by our heedlessness?

It was while I was struggling with this unhappy contradiction that I came upon Peter Russell's excellent book: *The Awakening Earth*[1] and so to the formidable scientific work of the Belgian physical chemist and Nobel Prize winner Ilya Prigogine.

Prigogine's theory of 'Dissipative Structures' throws much light on our current problems, and has evoked considerable excitement and discussion in many parts of the world, because it would appear that the chaotic mess in which society finds itself is exactly what one would expect from a living biological system in process of transition. The English edition of his book, entitled *Order Out of Chaos*[2] (co-author Isabelle Stengers) was published in Great Britain in 1984. Prigogine's book makes fascinating reading, even to the layman, and in the context of a living Earth its significance can hardly be over-emphasized, but it isn't easy. It concerns the behaviour of self-organizing chemical and biological systems known as 'open'systems, by reason of their interaction with their environment, and there is now reason to suppose that the planet on which we live, large though it is, may be one of them. For a while we thought of our planet as 'space-ship Earth', but it is more accurately described as an open, biological system in constant interaction with the solar environment to which we belong, and Prigogine's conclusions about such systems have very great relevance to us here on Earth.

Under certain conditions, described by Prigogine as "far from equilibrium", biological systems are found to have properties quite different from those of the mechanistic world view to which we are accustomed.

> A new type of order has appeared. We can speak of a new coherence, of a mechanism of 'communication'among molecules. But this type of communication can arise only in far-from-equilibrium conditions. It is quite interesting that such communication seems to be the rule in the world of biology. It may in fact be taken as the very basis of the definition of a biological system.[3]

Here we have three interrelated ideas, all of which were new to me. They are: 'closed and open systems'; 'communication among molecules'; and 'far-from-equilibrium'. All three are crucial to an understanding of Prigogine's work, and to my own, but they do take a little getting used to. Not only is Prigogine's usage of words highly technical, but familiar words like 'stability', 'order', and 'equilibrium' have a sneaky way of appearing in a sense quite contrary to expectation, so that apparently simple statements no longer seem to make sense. 'Stability', for instance, comfortable and reassuring in everyday life, and usually considered benign, is found to be wholly inappropriate to organic systems in the dictionary meaning of the term.

The dictionary meaning is 'steady and unwavering; resistant to fluctuation and change', which sounds fair enough. But organic stability actually *depends* on fluctuation and change, because a living system is in a continuous process of metabolism by which its cells are repaired, renewed or replaced. Capra notes that the stability of self-organizing systems 'is utterly dynamic and must not be confused with equilibrium'.[4] And this, for me, was another bewilderment. Surely equilibrium must infer a highly desirable state of balance between all facets of a living system? Once more I was wrong. According to the Oxford Dictionary, equilibrium is 'that state of a body in which the forces acting upon it are so arranged that their resultant at every point is zero'.

Now this is fine for chemical and mechanical systems, in which a state of equilibrium is one in which all processes involving energy have finally *stopped*, but it is clearly not applicable to a living system while it still lives. This explains why, in the discussion which follows, self-organizing living systems will be found to persist only when they are 'far from the state of thermodynamic equilibrium'.

'We now know', says Prigogine, 'that *far-from-equilibrium, new types of structure may originate spontaneously*.' (The italics are mine.)

> In far-from-equilibrium conditions we may have transformation from disorder, from thermal chaos, into order. New dynamic states of matter may originate, states that reflect the interaction of a given system with its surroundings. We have called these new structures '*dissipative structures*' to emphasize the constructive role of dissipative processes in their formation.[5]

In his book *The Turning Point* (1985) Fritjof Capra contrasts this situation with that of mechanical structures, citing a clockwork system as his example:

> a clockwork, for example, is a relatively isolated system that needs energy to run but does not necessarily need to interact with its environment to keep functioning. Like all isolated systems it will proceed according to the second law of thermodynamics, from order to disorder, until it has reached a state of equilibrium in which all processes - motion, heat exchange, and so on - have come to a standstill.[6]

This sounds simple enough, but I still wasn't happy. What worried me was why, when all energy had been dissipated into universally distributed heat, when all 'business' had ceased and calm prevailed, the result should be equated with disorder. At this point the sensible layman goes fishing!

The answer lies, of course, in the fact that all systems, both open and closed, are structured in a specific and characteristic way and their energies are ordered accordingly. The more complex the system, the more energy is required to maintain its structural and functional order. Thus the running down of available energy implies a state in which the system is progressively less able to maintain itself, and in a closed system, increasing entropy in terms of the *Second Law of Thermodynamics* is quite properly equated with a progression from order to disorder.

The Second Law of Thermodynamics is one of the cornerstones of physics. It concerns what is known as entropy. (The *First Law of Thermodynamics* concerns the conservation of energy. In a closed system energy can neither be created nor destroyed. It may change its form in a number of complicated ways, but no part of the totality of energy in a process is ever lost). Complementary to this law of the conservation of energy is that of increasing entropy - the *Second Law of Thermodynamics* - with which we are presently concerned. It describes a process by which less and less energy is available to do useful work, and by which, in the view of many scientists, the universe is gradually running down. Living organisms used to appear as an enigma because of their apparent defiance of this law. In the light of Prigogine's work on dissipative structures they assume a different role:

> Life no longer appears as an island of resistance against the Second Law of Thermodynamics ... it would appear now as a consequence of the general laws of physics ... which permit the flow of energy and matter to build and maintain functional and structural order in open systems.[7]

Dissipative structures, then, are self-ordering cross-catalytic chemical processes, so called because in an open system the entropy they produce is dissipated to the environment. Energy and matter are taken into the system in the form of external molecular structures, and broken down for use in the metabolism of the organism concerned. Since all living organisms - including man - are open systems in continuous interaction with their environment, Prigogine's work clearly has immediate bearing on our own functional processes, and, if Gaia is the discrete living entity some of us think she is, on the human social interactions which are part of her internal functioning.

Peter Russell notes that 'Phenomena such as the growth of plants, the regeneration of limbs in simple organisms, the excitation pattern of nerve cells, and many biochemical processes are now well understood as extensions of the principles found in dissipative structures.'[8] This concept has widened our general understanding of living systems, including the well-documented social behaviour - patterns of bees and ants. More recently it has been applied in studies of the brain, and has even been found applicable to human societies, to eco-systems and to the Earth itself.

I hope to show that this understanding of living things as 'dissipative structures' introduces a much needed note of optimism into current thinking. We are all painfully aware today that society is going through a period of major crisis. No-one escapes its repercussions, whether in an encounter with terrorism, in war, in the soaring cost of living, the theft of a motor car or the delinquency of a child - or simply in the horrendous pressures of earning the bread and butter a family needs.

Daily we listen to reports of devastation and death. Daily we learn of the depletion of resources, of pollution of the environment. Daily we are warned that the Earth is less and less able to renew herself, and that the very survival of humanity is at risk. Daily we read of the growing chaos in economic affairs. And as we move about our cities and parks, as we look in vain for beauty-spots free of litter, and people free of fear - as we gaze in dismay at beaches piled with debris and wild birds soaked with oil, it becomes increasingly clear that hand in hand with a usage of energy and matter that is manifestly not sustainable, humanity faces a frightening rate in the rise of entropy. In systemic terms these things add up to major fluctuations and instability in the system that the man in the street can no longer afford to ignore.

BUT ... and this is where the work of Prigogine and his colleagues is so immensely heartening - in considering the planet Earth as an open system on the brink of evolutionary transition to a dynamic new structural order, one salient characteristic of dissipative structures seems to stand out above all others. It is this: The emergence of a new structural and functional order actually seems to *demand* conditions of extreme non-equilibrium accompanied by severe fluctuation and instability.

If the energy or matter flowing through a dissipative system fluctuates, the organization will be retained as long as the fluctuations remain within certain limits. 'The system can even 'heal' itself',[9] says Peter Russell, through its self-organizing nature. If, however, the fluctuations increase beyond a certain limit, they will drive the system into instability. At this point there are two possibilities - collapse, or transition to a new level of organization. Thus it is the very condition of instability that makes new evolution possible.

> Characteristic of these transitions is a period of considerable chaos within the system. This instability corresponds to a maximum flow of energy and matter through the system, and maximum production of entropy within the system, requiring maximum dissipation of entropy into the environment. ...[10]

In the light of current ecological studies, this description has an ominously familiar ring, but there remains the encouraging side of the picture. We can either adapt, or be extinguished, like millions of other species before us. But we do have a choice. Evolution is no longer blind.

Russell points out that looking at humanity from the perspective of dissipative structures, we can see that two principal characteristics of a major fluctuation are present; increasing throughflow of energy and matter, combined with high entropy. We are now consuming energy and matter as never before in human history, while drastically increasing humanly induced entropy levels. Society reflects this trend towards disorder in a number of all too familiar ways, and social unrest,a rising crime rate, economic chaos and a general acute malaise have become the hallmark of our times.

Now the interesting thing is this: If we know all these things, and can see so clearly where we are heading, why don't we do something about it? If we know we are so near the breaking point, why sit by, feeling helpless, while civilization draws towards its end?

The single, heartrending, ridiculous, unbelievable answer is that we've let ourselves get bogged down in economics. All the things we could and should be doing take money we have not got. So most of the human world spends most of its available energy trying to make some more, till the scramble for money becomes an end in itself. A system that once spelled trade, development, and the spread of culture, has become an albatross around our necks. Profit is the force that rules our lives.

We are of course overwhelmed by the complexities, and by the sheer magnitude of the problems we seem to face, but I find myself asking just how far we can go on so foolish a path? What are the limits to suicide? And what are the limits to complexity, before we break?

Working with chemical and molecular systems, Prigogine notes that the more complex a system is, the more numerous are the types of fluctuations that threaten its stability. 'How then', he asks, 'can systems as complex as ecological or human organizations possibly exist? How do they manage to avoid permanent chaos?'[11] He sees partial answers to these questions in the stabilizing effect of communication, and of diffusion processes, noting that in complex systems, where species and individuals interact *in many different ways*, diffusion and communication among various parts of the system are likely to be efficient.

> There is competition between stabilization through communication, and instability through fluctuation. The outcome of that competition determines the threshold of stability.[12]

These discussions throw a penetrating new light on the condition of man in our century, and seem to me to offer a strong antidote to the understandable pessimism of most thinking people. We'll come back to them many times in the course of my own book, in which diffusion and communication are key issues. If we perceive the negative aspects of our society as a challenge rather than as the presage of doom, and if we apply our minds to meeting it by breaking out of the economic trap we are in, Earth may indeed take the final intuitive leap to a new order of consciousness in our time. To what heights she may then attain we can only guess, but we shall have played an exciting role in it.

PART THREE

Mind Emergence in Planet Earth

CHAPTER FIVE

3.1 First Principles

Summary: 'Mind' is perhaps built into the very fabric of the universe, and manifests at every level from molecule to man.

'I think the planet Earth is in process of becoming a contemplative entity possessed of nobler modes of consciousness than man, and that part of the process will entail an enhancement of awareness in individual men. And I think that personalization of the planet will trigger an exquisite communion of intangibles in the cosmos. Maybe its time I stopped thinking'.

Microcosm-Macrocosm, Heather Couzyn, 1971.

If we are like mites in this awesome new personage, can we, in part at least, be aware with its consciousness? In any living organism, how did its mind emerge?

Evolutionary theory suggest that life arose imperceptibly from inert matter when conditions in the primeval 'soup' were ripe for the chance formation of organic molecules. After that first crucial happening, the theory goes, the complex organization of molecules gave rise to living cells, whose gradual organization into more and more complex units finally made possible the primitive brain. Inert matter had given rise to mind.

I can't quite see it that way, because nothing, to me, feels inert. In its own way, everything I touch seems aware. Something flows between us. It answers back; responds. We are kin. We sprang, after all, from the same cosmic impulse, the same creative force, the same God. We've come the same long journey together since Time began. We become aware of each other through sight or sound, through smell or vibration or heat, but most of all through touch, the primary sensory mode, I believe, of so-called inanimate forms. After years of obsessive contemplation of these mysteries, I am left with the sense of an all-pervading intelligence extending itself in ever more complex areas of experience, with the whole of it expressed, as in a hologram, in the smallest and simplest part. And just as the activity of any subatomic particle instantly affects every other with which it has been in contact, so must mental experience maintain a continuous whole. Mental activity of any part of the Earth must instantly pervade all of it. Thus the whole mental experience of the planet, from molecule to man, maintains unbroken continuity.

Mind has been described by David and Doris Jonas as an 'emergent quality ... a product of the complex interaction of the billions of individual neural cells, which does not reside in the properties of any one of them.' In their fascinating book *Other Senses, Other Worlds* (1976) they suggest that in some way not understood, thought is a quality that emerges from the sheer number of interacting nerve cells, and a single thought can be conceived of as their 'multiple echo'. [1]

This truly marvellous evocative phrase stirred my thought processes more keenly than most I had met at the time of reading this book. I had battled so hard over more than thirty years to picture the result of the trillions of interacting and super-imposed electro-magnetic fields of all forms of life on the planet, and the Jonases, in one inspired phrase, seemed to say it all. They were, however, at pains to point out that this conceptualization only described, and did not explain, the emergence of an intangible thought from a tangible mass of living matter, which gave me to wonder, even then, if it were not indeed the other way round - a conjecture that has since become conviction. If the primary creative principle of the universe ('God' is the only word we know for anything so awesome) gave rise to an intelligent system of energy, which in turn precipitated matter, then not only 'life', but even inanimate matter has its own mode of intelligence, innate and of its very nature.

We like to think of ourselves as the torch-bearers of consciousness, but is there any reason why a star, for instance, might not have its own mode of being aware, its own response to the creative impulse, its own system of dynamics in the cosmic thought process? Is our instinctive negation of so far-fetched a notion only another of our chauvinisms?

We've been led to think of evolution as progress from inanimate to animate matter, and thereafter to ever more complex forms of life and thought. We marvel that out of the wasteland of inanimate matter, life and intelligence arose. But if mind preceded matter, then all the lovely intangibles, all the lovely forms are facets of the same diamond. All are functions of the primary creative principle, undefined, and, in our terms, probably indefinable.

This sort of concept has as its starting point the possibility of a precreation spiritual cosmos whose primary principle could bring into being the intelligent system of energy we think of as life.

If, then, this principle begets the force from which all other energies spring, and of which they are no more than different states, then thought, or at least the potential for thought, is built-in to the very fabric of the universe, animate and inanimate: every change in the matrix will affect us: every thought we think will have immediate effect on the primary energy patterns of the universe, and particularly on the electro-magnetic energy fields of our cells, our bodies and our environment, right down to the basic atoms and electrons from which we are built. If the primary force is holy, all the world embodies its holiness. Had we no other standard for human ethics, then this, of itself, would be basis enough. We have need for respect for everything there is.

CHAPTER SIX

3.2 'Mind' in Microscopic Life Forms

Summary: Intelligent behaviour is observed in primitive single-celled organisms such as amoeba.

If intelligent energy informs even inanimate matter, what is it that distinguishes 'life' from non-life? The mystery persists, however far you probe. There is of course the fact that even the simplest cell is a system, displaying clear co-operative tendencies. Not only is it extremely complex, but its parts are inter-related and inter-dependent to the point of exhibiting a clearly definable life cycle, a progressive series of changes from germ cell to death of the organism. It is self-organizing in the sense that it maintains a high degree of internal homoeostasis; it exhibits the entropy-reduction mentioned above, and it is self-reproducing. It takes in matter, energy and information, processes them in accordance with the needs of its particular metabolism, and excretes the waste to the environment. Inter alia it is interesting to note that this excreta, 'waste' to the system that excretes it, is usually of value to another living system, or to the general function of the living environment in which it finds itself. Brian Ford makes this point delightfully in the context of the yeast of fermentation, (*Saccharomyces cerevisiae*).

'In still wine', says Ford, 'the yeast's by-product of alcohol is accumulated as the carbon dioxide bubbles; in bread it is the carbon dioxide which is trapped and which causes the bread to rise, while the alcohol evaporates.'[1] *Yet both these products are waste to the yeast.*

> 'It is not an overstatement to say that alcohol and the carbon dioxide are the excrement and the exhaled breath of the living microbe.'[2]

This is an entertaining side-light on the mystique of wine-making, but worth remembering, for the principle of reciprocal waste-usage has important applications for industry. We'll come back to these towards the end of this book when discussing the application of natural principles in economics.

None of the activities of living systems, taken alone, can be regarded as *definitions* of life, but where they all appear together in dynamic interaction, we have little difficulty in *recognizing* the presence of life.

Digging a little deeper still we are struck by the enormous information content of even the simplest living cell. We had to wait for cybernetics and computer technology before we could quantify this, but what emerged was a difference of many orders of magnitude between the information content of, say, a crystal of metallic ore, and the genetic content of even the simplest living cell. This difference in information-content seems to run parallel with the evolutionary complexity of 'living' *vis à vis* 'non-living' systems, and concurrently with the intelligence level of the evolving brain. Most of us are comfortable with this notion. Leaving aside the enigmatic possibilities of whales and dolphins, it places us way at the top of those living forms with which we are familiar. But today we are able to be more precise than this. The likelihood of mental activity can now be quantified by the minimum information content in the organism's genetic material.

A typical DNA molecule in a human being is composed of about five billion pairs of nucleotides, which in computer terminology can be expressed as twenty billion (2×10^{10}) 'bits' of information. Simpler organisms have less to do and therefore require smaller amounts of genetic information. Carl Sagan points out that the landers that put down on Mars in 1976 each had pre-programmed instructions in their computers amounting to a few million bits.

> Thus Viking had slightly more 'genetic information' than a bacterium, but significantly less than an alga.[3]

To the layman such a comparison is startling to say the least. We tend to be overawed by man's accomplishments and to have remarkably little understanding of the awesome capabilities of the lowlier forms of life. Yet the free-living single-celled protozoa can carry out many of the basic activities of higher animals. Here is Brian Ford again. Speaking of the ciliate *Epidinium ecaudatum* found in the rumen of sheep and cattle, he says:

> So here we have microscopic forms of life which, *within one cell,* have organelles of movement, mouths, muscles, even sometimes a skeleton of a sort; they may have a kind of nervous system, perhaps a 'brain', and their relatives (e.g. the alga *Chlamydomonas*) can see where they are going through eyes that are clearly recognizable as being built on the same pattern as the human eye ...[4]

The information content in the DNA of *Epidinium ecaudatum* is clearly very high, so let us look at a less clearly structured protozoan, the familiar amoeba. Brian Ford compares it to 'a spoonful of treacle running down a plate.'[5] There are no organized features within this cell. Amoeba was described in 1870 as 'a simple mass of organic matter'; and, in 1968 as 'a naked bit of protoplasm.'[6] Yet Ford points out that this bit of naked protosplasm is immensely complex and very different from the blob of treacle on a plate. Amoeba can move where it likes, uphill as readily as down, and systematically projects itself. 'In what way,' asks Ford, 'can a predominantly featureless jelly undertake to contract, to flow, to undergo countless predetermined changes in viscosity, to circulate its protoplasm, and so to move?'[7] To this I would add, in open-mouthed wonder: How does this blob of naked protoplasm evolve the capacity to discriminate between the varied stimuli surrounding it, in such a way as to know in which direction it would 'like' to go?

Molecular biology has come a long way since Ford's book was published in 1976. Speaking of bacterial chemotaxis, a process in which stimulus takes the form of a gradient of chemical concentration, Prigogine remarks:

> We know of approximately twenty or thirty receptors that can detect highly specific classes of compounds and make bacterium swim up spatial gradients of attractants or down gradients of repellents. This 'behaviour' is determined by the output of the processing system - that is, the switching on or off of a tumble that generates a change in the bacterium's direction.[8]

Knowing a little more of how these processes work does nothing to reduce the marvel of their complexity. It even increases it. The so-called 'simple' cells of the lower forms of life get less and less simple the deeper we explore.

Ford says that if you watch the way the granules within the cells move, you can see that amoeba has a preferred orientation, so that one part of the cell always leads the way.

> But how can an irregular drop of cell substance know which part of it is its 'head'? How does the cell keep itself intact, when it exists in the form of soluble complexes dissolved in water, yet living in water itself?[9]

These organisms are able to creep towards a favourable stimulus, temperature, or oxygen concentration. 'How,' asks Ford again, 'can they do this?'

One could argue that the sense of direction is actually one of memory, so that the cell compares the amount of favourable stimulus at any given

> time with what was there before, and so keeps moving 'uphill' towards the target. But does that answer the question? Only if we admit that we cannot imagine how Amoeba could possibly aspire to the possession of memory. Amoeba has no component cells and therefore not brain.[10]

When you consider that Ford is discussing one of the most primitive life forms known, it is clear that there is an enormous gulf between the information content of a single-celled protozoan and that of any inorganic chemical compound - so great a gulf, indeed, that only the fusion of micro-biology and computer science could make possible the formidable mathematical analysis necessary for an understanding of the genetic content of the living cell.

CHAPTER SEVEN

3.3 Human and Global Mind

Summary: The human brain is discussed as a possible 'neuronal' component of Global Mind.

At the other end of the evolutionary scale, or so we like to think, is the human brain, a collection of interconnected neurons believed to be the active elements in brain function. At one time it was believed that these neurons behaved as simple switching elements similar to those in an electronic computer, and that the enormous information content of the human brain was due solely to the complexity of interaction between them. Later it became clear that there were electrical micro-circuits in the brain which made possible a much wider range of responses than those of the switching elements in a computer. But even this was a gross oversimplification. When you remember that all the cells in our bodies, including those of our brains, have been derived by division from the original fertilized egg carrying our genetic inheritance, and when you add to this that a typical DNA molecule (just one molecule!) in such an egg carries about 20 billion 'bits'of information, it is clear that to conceive of the neuron as a simple switching element is to underestimate the complexity of the human brain to the point of absurdity. In Brian Ford's words:

> If that is all neurons do, then they would have to be the most primitive and unenterprising cells in our bodies.[1]

Allowing for the suppression of certain of their functions by specialization and the enhancement of specifically 'cerebral' functions of others during the same process, the mental activity of human brain cells must surely outdo those of the primitive amoeba by many orders of magnitude, and the complexities of human brain function take on even more staggering dimensions than before.

We have been talking about what happens within the brain of a single human being, but in the global mind posited for the future, each such brain would behave as a single cell: ten billion human 'neurons' in corporate cerebral functioning; ten billion interrelating humans in a global mind, whose capacities would be enhanced by the almost incomprehensible complexities of all the single human brains within it. Gaia, when she comes to her own, should find herself possessed of superb mental equipment, and concurrently with a consciousness difficult to imagine at the present time.

If then, the Earth is, or is becoming, a single conscious living organism, and if individual men can be thought of as single neurons in its evolving brain, then the process of multicellular specialization will have come full circle, in the sense that the life-force will once more have assumed the form of a single-celled organism (no longer a contradiction in terms) with all the enhanced richness and diversity of function of its multicellular experience. In the cosmic context it could be regarded as a purposeful microbe, and a fine field of enquiry opens up concerning its relationship with its cosmic environment.

For practical purposes, though, we are concerned with the nature of man in relation to planet Earth, and in the immediate effect which a living, conscious Earth would have on our everyday lives. So what is crucial is the potential of the human brain as a unit in planetary intelligence, and the ways in which individual human beings might interact together to generate a wider emergent consciousness.

The cognitive areas of the human brain appear to contain about ten billion switching elements or neurons. Carl Sagan points out[2] that if each human brain had only one synapse, we would be capable of only two mental states. But the human brain contains some 10^{13} synapses.[3] Thus the number of different mental states of a human brain (discounting for the moment the microcircuits and controversial intraneuronal function of individual cells), is at least 2 raised to this power *i.e. multiplied by itself ten trillion times*. This is an unimaginably large number, greater by far than all the electrons and protons in the universe. It follows that for all practical purposes the possible mental states of a single human being are all but infinite.

If then, you envisage the evolution of planet Earth as a movement toward maturity in a single self-conscious living cell, and if you conceive of its 'brain' as embracing the whole of mankind, then the potential for its states of consciousness becomes enthralling indeed. Add to this the possibility discussed above that in each individual man a single neuron may perhaps carry on certain integrative processes of thought and cognition within its own structure, and the possible states of consciousness for planet Earth are past imagining.

When you come back from dreaming and really *look* at our world, the dislocation is insupportable. Your heart lurches sickeningly to your boots. You are forced to wonder if fusion of all that hatred could possibly lead to anything good. Will Gaia be a monster, or divine? How can anything good evolve from so much cruelty and greed?

The answer is, of course, quite simple. *The process of symbiosis is benign.* It has to be. There is no other way that it could work. Gaia, if she survives at all, will be beautiful.

PART FOUR

The Cosmic Egg

CHAPTER EIGHT

4.1 Genesis

Summary: Planet Earth may have been seeded from the Universe, coded like any other living thing.

If Earth is a microbe, an egg, or an embryo, as I take her to be, she must have been seeded from the universe, coded like any other living thing. Her contours would gradually have become discernible as rocks and sea, and evolution would have to be seen as tissue differentiation in the developing embryo.

When I first thought of this it startled me, though I should be used, by now, to my wild ideas. I set it aside to incubate, but it wouldn't go away. The egg refused to be simple or explainable except in living terms. From an egg could come a spider, a hen, or a crocodile; and from an egg came a man. Born of a coded microspeck, he would bestride the universe. Nothing I thought of was as astounding as fact. Yet the thing was getting a bit out of hand. A cosmic egg? It was too much to digest. I couldn't take it seriously. Now and then, tongue in cheek, I considered it.

Then in 1981, I saw a related idea in print in Hoyle and Wickramasinghe's book *Evolution from Space* (1981).[1] It seemed that the simple biological beginnings we'd been looking for on earth did not exist, since most of life's biochemical complexity was already present at the time the earliest surface rocks were formed. Could life have come in from space? These two authors thought it had. The elements of life, in their view, had been seeded continuously from the universe over a very long period of time, and this process still continued today.

To make the issue still more bewildering, Prigogine, writing in 1984, cites the same data - the early appearance of life on earth - as an argument in favour of the idea that life is the result of spontaneous self-organization that occurs whenever conditions for it permit.

> We are tempted to go so far as to say that once the conditions for self-organization are satisfied, life becomes as predictable ... as a falling stone.[2]

These two points of view are not necessarily in conflict, since such conditions could conceivably arise both in space and on the earth itself. But suddenly, and quite illogically, since this is not what any of these men were saying, my own notion of the egg re-emerged. To be sure life was seeded from the universe, but the seeding began before the Earth was formed. We are a speck of dust in the universe and are part of it. A living impulse had coded and nurtured us. In the swirling clouds of gas before we were formed, it had given us life.

You'd think this would be enough for a while, but it wasn't to be. The moment a thing feels certain, its interest palls. One has to move on. 'So what?' I wondered. 'How does this help us down here, in human life? Can it feed our starving millions or ease our pain, or erase the hate from our hearts?' I knew I'd have to try and answer this in the end. I didn't know how. I put it off as I had with the notion of the egg. It was too big for me. But it wouldn't go away. Years later, towards the end of this book, I was to tackle it again.

Meanwhile the sense of urgency mounts. For the first time in human history all our basic disciplines begin to converge, and attention focuses on the plight of mankind. From a thousand different angles at once, people are asking important questions, not just because of their academic excitement and worth, but because of our perceived and desperate need. All seem to point in the same direction - to the unity of life and of universe. In Stephen Hawking's words:

> This large-scale homogeneity of universe makes it very difficult to hold an anthropocentric view or to believe that the structure of the universe is determined by anything so peripheral as some complicated molecular structures on a minor planet orbiting a very average star in the outer suburbs of a fairly typical spiral galaxy.[3]

Insignificant though we be, we are clearly very much part of the universe.

My own reaction is this:

Can we discover a cosmic pattern useful to man? Is any such pattern observable in the systems we know? Does it reveal itself in the behaviour of elementary particles? in the molecule? in the cell? Can it be seen at work in nature? Can it be extended to ourselves? And most important of all, can it *help*?

Until quite recently it would have seemed there were insurmountable barriers to this kind of thinking. We were told that the mathematical laws of Probability Theory forbade the valid transfer of concept between the very small and the very large. However, just as matter travelling near to the velocity of light defies the 'laws' of Newtonian physics, so may the relation between microscopic and macroscopic levels be altered when a system is in a state of non-equilibrium such as Prigogine describes.

Heisenberg showed that we can never know, simultaneously, both the position and the momentum of an atomic particle. The behaviour and lifespan of a single particle is entirely unpredictable. But Prigogine takes this observation a step further, asserting that under certain conditions, randomness remains essential for very large living systems as well. In Prigogine's view, the role of 'fluctuations' in 'nonequilibrium systems' shows that some of the properties of quantum mechanics (discovered on the microscopic level), now appear at the macroscopic level as well. Under certain conditions the very small and the very large behave in the same way.[4]

This observation is extremely important for my own work, because, in my search for understanding, I rely intuitively on basic principles that run like a refrain through all of nature, and look for ways in which their most successful patterns throw light on our problems down here.

Indeed the greater part of my work rests on the application in social and economic affairs of phenomena observed *within the molecule*. So I'm glad to know that whatever the principles on which these self-organizing microscopic systems operate, we can expect to find them, under certain conditions, at all levels of life. It follows that should a discovery in the field of molecular or cellular behaviour appear to be useful, it need not necessarily be out of court as a guide to human affairs.

As I write, Halley's comet is streaming past the Earth, for all the world like some kind of cosmic sperm. Ridiculous fancy, like little green men on Mars - that kind of thing - but it pleases me. The coding of cosmic sperm in a microdot! Starring the velvet sky like bright-blown pollen, seeds of the universe blinding us with light.

CHAPTER NINE

4.2 Inside the Molecule

Summary: Microscopic Cybernetic Systems suggest a model for Society.

I found what I was looking for at last in two paragraphs on microscopic cybernetics in the work of the Belgian molecular biologist and Nobel Laureate, Jacques Monod. His book, *Chance and Necessity*, was published by Collins in London in 1972.

Even though I'd been waiting for it, I was stunned. It came out at me from the printed page like a bolt of lightning, with a conviction of truth as certain as it was strange, for I didn't understand a word of it with my intellect. That would come later. At first it was just excitement - an overwhelming sense that I held in my hand the key to everything I'd been searching for through thirty years. All I had to do was to understand. Then I would know what this new knowledge could mean to the future of man. Had I known at the time how long understanding would take, I might have thrown in the sponge. It took me all of ten years to see where it led.

The short quotations from Monod's book that wrought such a turmoil in me will be found in the Appendix. Only someone trained in a related discipline will appreciate fully the niceties of scientific meaning involved. Nevertheless, in the context of my own work, they give birth to some very important ideas.

They are quite the most significant thing in this book.

The first paragraph suggests that functions we have been accustomed to equate solely with the cerebral performance of so-called 'higher' intelligence, and especially with the thought processes of human beings, are by no means our exclusive preserve. They are observable in living cells *at the level of the protein molecule.*

In the second paragraph we learn that these microscopic molecular systems appear to reveal not only the capacity to 'choose' a reaction, but '*consistent criteria for such a choice*'. And, most surprising of all (and most humbling), the sole criteria of choice is '*their contribution to the coherence of the system*'. [1]

To a species accustomed to regarding itself as the sole custodian of complex cerebral performance, this is an astounding revelation. Absolutely stunning really, when you consider the context, for we are talking here of the interaction of atoms! - the *atoms* which make up protein molecules within a single cell!

Monod states that the way in which these microscopic cybernetic systems work 'permits a complete freedom' in the choice of controls. This freedom gives molecular evolution 'a practically limitless field for exploration and experiment', enabling it to develop the spontaneous network of cybernetic interconnections on which an organism operates, and by which it is enabled to 'transcend, if not to escape, the laws of chemistry'.[2]

In the field of molecular biology Monod's work could hardly be more authoritative, nor his contribution to evolutionary genetics more significant. Hardly a book one picks up in any related field is without reference to his research, but I must hasten to add that the philosophical conclusions presented in my own book are entirely mine, and are startlingly different in mood from those drawn by Monod himself. Monod seems to be left with a sense of desolation; even of alienation.

> 'Man must at last awake ... to his total solitude, his fundamental isolation. Now does he at last realize that, like a gypsy, he lives on the boundary of an alien world. A world that is deaf to his music, just as indifferent to his hopes as it is to his suffering or his crimes.'[3]

This saddens me. But not all Monod's fellow scientists share his pessimism. Prigogine observes:

> This is a paradox. A brilliant breakthrough in molecular biology, ... ends upon a tragic note.[4]

To anyone like myself, dedicated to seeking coherence in human society, Monod's results could hardly be more exciting, whatever the philosophic conclusions he draws from them. As already explained, I have no knowledge of molecular biology, in which I am completely out of my depth, and I hope Monod will forgive me for using his work out of context so shamelessly. My only excuse is the flame his words kindle in me, for these revelations of natural law could perhaps be applied to mankind. Our bewildering social complexity gives us, too, a practically limitless field for exploration and experiment. If these complexities, by their very variety and freedom of choice, can enable mankind (and Gaia) to elaborate her cybernetic connections (as happens within the molecule) then we have at our disposal not a despairing disarray of problems, but the natural groundwork of our salvation.

Monod concludes that:

> ... the study of these microscopic systems shows us that for complexity, for richness, for potency, the cybernetic network of living beings far surpasses anything that might be predicted by the study of the overall behaviour of whole organisms. And though these analyses are still far from furnishing us with a complete description of the cybernetic system of the simplest cell, they tell us that, without exception, all activities that contribute to the growth and multiplication of that cell are interconnected and intercontrolled, directly or otherwise.[5]

Monod's observations excited me so much that they became the 'Leit Motif' of the second part of this book. Here was a way to translate concept to policy. If the key factor in all natural cybernetic systems was their reliance on *freedom of choice,* right down to the molecular and atomic level, then this freedom began to appear as a fundamental law of life, and could be applied as such. Most important of all, if the regulatory interactions of cybernetic systems succeeded in being chosen '*solely for their contribution to the coherence of the system*', they might illumine for us the basis of a new human ethic. There might quite possibly be no more fruitful way of solving some of our global problems than by thinking these astounding revelations through to the very limit of their applicability in human affairs.

CHAPTER TEN

4.3 Inside the Cell

Summary: Microscopic animals are found to have maintained their separate identity within the human cell, over millions of years, while fulfilling functions vital to that cell.

In a ferment of excitement at these revelations of natural law, I moved on to look at the cell. Did the same principles hold at this yet more complex level? Compared with even the largest of the protein molecules, the cell was a veritable universe.

I knew that molecular biology had made spectacular advances in the last decade or so. I also knew that its intricacies would be way above my head. My knowledge of biology was a random mix of standard biology texts at schoolboy level, long outdated, general reading of everything I could lay my hands on, and one or two 'popular' presentations designed specifically for the inquisitive layman.

In 1984 Prigogine was still able to say that the problem of biological order involving the transition from molecular activity to the super-molecular order of the cell was still far from being solved,[1] so I wasn't going to find any ready-made answers here. But dating back to 1974 and shining like a beacon in the context of my own work, was Lewis Thomas' book *The Lives of a Cell.* [2] As a popular presentation of cellular biology, I've come across nothing else to touch it.

> Man is embedded in nature ... The new, hard problem will be to cope with the dawning, intensifying realization of just how interlocked we are. The old, clung-to notions most of us have held about our special lordship are being deeply undermined ... We are not made up, as we had always supposed, of successively enriched pockets of our own parts. We are shared, rented, occupied.[3, 4]

I learned that our cells harbour, and indeed are made up of a number of independent entities, among them mitochondria, centrioles, basal bodies and others, without which they would be unable to function, but which are in effect really just getting on with their own independent lives. They turn out to be little separate creatures, the 'colonial' posterity of primitive migrant bacteria. They swam into the ancestors of our own cells many millions of years ago and stayed there. Ever since, they have got on with their lives in their own special ways, replicating in their own fashion inside our cells. They have their own DNA and RNA (which are quite different from ours). Without them, says Thomas, 'we would not move a muscle, drum a finger, think a thought.'[5] But they have maintained their own identity. Our cells are no longer the 'pure line entities' we learned about at school; 'they are ecosystems more complex than Jamaica Bay'.[6]

Over the last few million years these microscopic little animals have somehow learned to act in concert in such a way as not to destroy the living human cell which is their habitat. They seem to have learned (without any help from us!) about symbiosis - a lesson we have still to learn.

This linking of microscopic ancestral creatures in symbiosis, within a single functioning cell, is a marvel of cooperative accomodation. But the process is a general one in nature, extending to the plant world too. Even chloroplasts, which provide food and oxygen for the rest of us through photosynthesis, turn out to be separate little creatures with their own genomes. But so efficient is their coordination in the green leaf, that the global task they perform goes almost unnoticed.

No-one is sure how creatures like the primitive mitochondria got into our cells in the first place, but however it was, these microscopic beings within our flesh have maintained their separate identity for more than a billion years, in spite of the intimate symbiotic arrangements of their tenancy. It is therefore perfectly possible for infinite numbers of discrete self-motivated individuals to interact symbiotically, thereby maintaining the functional efficiency of a larger whole. When one tries to stretch one's imagination to envisage the *modus operandi* of such processes, it is impossible not to conclude that, in the cell as in the molecule, their interactions '*succeed in being chosen solely for their contribution to the coherence of the system*'.[7]

If Gaia is in truth in process of convergence into a living 'cell', we human beings still have much to learn!

So much, then, for the cell. But cells form themselves into organs, organs to people, people perhaps into this Supercell, this Earth, this embryo. Moving from microscopic to macroscopic, do these same principles apply? Lovelock's thoughts on Gaian cybernetic systems would seem to suggest that they do. He likens the complex interplay of global phenomena such as temperature control and salinity to bodily regulatory function, noting ruefully that disentangling the constituent loops of such a system is unlikely to be easy:

Salinity control, for example, may be a key Gaian regulatory function. If its details are as intricate and complex as those of that amazing organ the kidney, then our quest will be a long one. We now know that the kidney, like the brain, is an information processing organ. To achieve its aim of regulating the salinity of our blood, it purposefully segregates individual atoms. In every second it recognizes and selects or rejects countless billions of atomic ions. [8]

Here once again, this time in an organ of the human body, we find the selective principle at work. The intricacy and innate intelligence of life at all levels cuts the breath, but if we project the same principles into a hypothetical planetary entity, one key question predominates:

If coherence of the system is the only law, in Gaian as in molecular evolution, and if individual freedom of choice acts to bring this coherence to fruition, how, from moment to moment, are we to make that choice? How does one orientate oneself towards the ultimate good? How does one *know*, in Gaian terms, which of one's thoughts and actions contribute to her coherence?

One does not know. One turns toward it blindly, like turning to God. And indeed this turning is what it is all about. We are feeling our way.

Our need is the self-organization of ten billion independent minds into a corporate will. Nothing is pre-determined. Nothing is sure. At every point we are confronted with choice. Uncertainty opens before us like branching twigs on a tree, at every fork a bewilderment. There is no going back. We teeter on a knife-edge of danger and have to choose. But if there is danger, there is also hope. Such is our instability, the smallest movement can be critical, one man tipping the balance for good or ill: for chaos, for coherence or for death.

So much I knew. But in such a predicament, how was a man to proceed? Could a balance be achieved between free choice and necessity such that the will of ten billion individuals should incline us towards coherence and to life? It could only be achieved, I thought, if each man carried within him unconsciously, an archetypal imprint of a whole - only, that is to say, if coherence were the norm from which he sprang. His choice would then embrace this wider truth.

Was there such an archetype for man? for Gaia? Was there an archetype for everything? It seemed my most urgent need to know, and to know how it worked. Did anyone know? Did anyone understand the emergence of forms? Why the wild winged bird or the whale, and why the rose? Somehow I had to find out.

CHAPTER ELEVEN

4.4 The Search for Forms

Summary: The recurring structures of living forms are perceived to be partly ordered within non-energetic spacial fields known as 'morphogenetic fields'.

Earth may conceivably have developed within such a field.

This time it was Rupert Sheldrake's book *A New Science of Life* (1981)[1] that came to my aid. In it Sheldrake offered a clear-cut hypothesis of the development of form and structure, based on a concept he calls 'morphogenetic fields'. It was all so new to me that it took a while to sink in, but its relevance to my own work was clear enough. Morphogenetic fields as described by Sheldrake have clear affinities with the archetypal imprint I'd been searching for.

My own search had originally been directed at recognition of those principles in nature which had given rise to beautiful and functional forms. I knew that evolution tended to repeat successful experiments in many different contexts, with the result that the same patterns might appear over and over again in forms as different as a shell and the petals of a flower. So at that time I had wondered if there were any of nature's structures fluid and adaptable enough to be useful as a model for society. How, I had wondered, did nature succeed in grouping cells together in so stunning a variety of workable forms?

Later, as the notion of an almost genetic 'seeding' of the Earth took hold, my search for structures gave way to thoughts about a possible inherent form, within which the Earth might be growing, in the same way that a man would grow from a fertilized egg in the womb. Human genetic processes were clearly not applicable, and I had begun to ponder in terms of energy fields which might in some way reflect universal patterns. If Earth's potential (including our own) was in some way innate in the coding of an embryo seeded from the universe, we had only to become attentive, tune in to its impulses, and direct our energies to furthering its greater good. The fields of force inherent within it would then direct the enterprise.

Energy fields are familiar to most of us in today's technological world, and I knew that living bodies also created a measurable electro-magnetic field around them, so at one time I had wondered whether this life-field might have anything to do with the creation of form in developing embryos. Every child has succumbed to the charm of watching iron filings assume a pattern in the electromagnetic field of a magnet. So, conceivably, could ions, atoms, molecules, cells, or organisms conform to the life-field of the developing system - always supposing, of course, that such a field pre-existed in the developing embryo.

There does seem to be some evidence that this is indeed the case. Lyall Watson quotes Burr as having introduced micro-electrodes into a newly laid frog egg,[2] and having found that, even before the egg began to divide and develop into a tadpole, he could measure voltage differences in those parts of the egg that were due to become the nervous system. The egg material that would eventually serve the function of communication was already displaying the voltage characteristic of that part of the organism.

Sheldrake extends the idea of electro-magnetic fields to that of *non-energetic* spatial fields, which he conceives as having a causative role in the repetitive emergence of form. These 'morphogenetic' fields are seen as exerting an influence by virtue of the form in question having *existed before*, and Sheldrake makes no attempt to define the origins of such forms. He suggests that although energy can be regarded as the cause of change, the *ordering* of change depends on the non-energetic spatial structure of these fields. To a simple soul like myself it is a short step from this non-energetic-spacial-field-structure to the concept of thought as the formative agent in patterns of growth. These morphogenetic fields could be thought of as functions of the 'mind'aspect of the Primary Principle - or, quite simply, as thoughts in the mind of God.

Sheldrake wisely leaves such questions wide open, and confines himself to the concept of morphogenetic fields as an extension of known energy fields. He points out that gravitational and electro-magnetic fields are spatial structures which are invisible, intangible, inaudible, tasteless and odourless; they are detectable only through their respective gravitational and electro-magnetic effects.

In order to account for the fact that these physical systems influence each other at a distance without any apparent connection between them, these hypothetical fields are endowed with the property of traversing empty space, or even actually constituting it. Sheldrake points out that in one sense, they are non-material; but in another sense they are aspects of matter because they can only be known through their effects on material systems. In effect, he suggests that the scientific definition of matter has simply been widened to take them into account. Similarly, morphogenetic fields are spatial structures detectable only through their morphogenetic effects on material systems; they too can be regarded as aspects of matter, if the definition of matter is widened still further to include them.[3]

Sheldrake includes in his hypothesis of formative causation both biological and physical systems at all levels of complexity:

> Since each kind of system has its own characteristic form, each must have a specific kind of morphogenetic field: thus there must be one kind of morphogenetic field for protons; another for nitrogen atoms; another for water molecules; another for sodium chloride crystals; another for the muscle cells of earthworms; another for the kidneys of sheep; another for elephants; another for beech trees; and so on.[4]

If this is so, I wondered, could there not be just such another for the whole, stupendous Earth cell, and another for the universe to which we belonged?

Sheldrake explains the process whereby the forms of previous systems might influence subsequent similar systems, by comparing it with the physical phenomenon of resonance.

> Energetic resonance occurs when a system is acted upon by an alternating force which coincides with its natural frequency of vibration.[5]

We are all pretty familiar with this, using it daily when we tune in our radios to broadcast frequencies, or when a stringed instrument resonates to appropriate sound waves. Its significance lies in the selectivity with which a system responds only to its own frequencies, but while this kind of resonance involves energy, the resonance of form that Sheldrake talks about (morphic resonance) is non-energetic.

According to Sheldrake's hypothesis, morphic resonance causes the form of a system, including its three-dimensional internal structure and vibrational frequencies, to become *present* to a subsequent system with a similar form. So rose becomes present to a later rose; bird to a bird; or whale to a whale, thereby influencing the development of form of each. The pattern of the former superimposes itself on the latter. And the influence works to a lesser extent in reverse as well. Not only does a specific morphogenetic field influence the form of a system, but also the form of this system influences the morphogenetic field, and through it becomes present to subsequent similar systems.

I like this feedback principle in the hypothesis of formative causation, because it fits with my own perception that everything influences everything everywhere in the universe. This means that every entity, whether man, molecule, or Gaia herself, would have a creative role from moment to moment in cosmic as well as in local evolution. It also ties the Earth very specifically to earlier universal patterns to which we may be related.

Sheldrake suggests that since morphic resonance is non-energetic, and morphogenetic fields themselves are neither a type of mass nor of energy, there is no *a priori* reason why it should obey the laws that have been found to apply to the movement of bodies, particles and waves. It could be just as effective over ten thousand miles as over a yard, and over a century as over an hour.

In terms of this thesis, when a morphogenetic germ comes into morphic resonance with the forms of countless previous higher-level systems, these forms do not coincide exactly, but give rise to a probability structure. As the first stages take place in the genesis of form, structures are actualized at particular places within the regions given by the probability structure. The system now has a more developed and better defined form, and will consequently resemble the forms of some previous similar systems more closely than others; the morphic resonance of these forms will be more specific and hence more effective.

If the cosmic coding of the Earth cell which I envisage has truth in it, then the embryo could well work along these lines, and we are more closely related to certain earlier cosmic life systems than astronomy has so far been able to ascertain.

Sheldrake suggests that the morphic unit is continuously stabilized by its morphogenetic field. In biological systems, this maintenance of forms enables the morphic units to persist, even though their constituent parts change as they are replaced. So the 'I' remains 'I', even though millions of my cells are replaced each day. This effect is further enhanced in Sheldrake's view by the fact that this morphic resonance will include a contribution from past states of the system itself:

> In so far as a system resembles itself in the past more closely than it resembles any other past system, this self-resonance will be highly specific. It may in fact be of the most fundamental importance in maintaining the very identity of the system.[6]

This observation seems to me to explain in some measure the extraordinary persistence of identity in living forms at all levels of complexity - such for instance as the 'intracellular lodgers' in our 'intelligent flesh' already mentioned. It also reinforces even further the likelihood that the personal identity of men will persist after the fusion of intellect in the Gaian mind. Perhaps this is one of the meanings of 'immortality' so dear to the human soul. I do not think it conflicts with holiness.

CHAPTER TWELVE

4.5 The Collective Unconscious

Summary: Carl Gustav Jung postulated a common substratum of consciousness shared by all human minds.

This idea is extended to posit a global sub-conscious, related to the archetypal morphogenetic field of mankind.

Carl Gustav Jung tried to explain the inheritance of the collective unconscious mind of man by suggesting that the archetypal forms were 'present in the germ plasm.'[1]

Sheldrake feels that it is very doubtful that anything with the properties of the archetypal forms could be inherited chemically in the structure of DNA, or in any other physical or chemical structure in sperm or egg cells, but observes that mental phenomena need not necessarily depend on physical laws, but rather follow laws of their own.

> The Mind's properties could be such that past mental states are capable of influencing present states directly, in a manner that does not depend on the storage of physical memory traces ... [2]

In this hypothesis, if memories are not stored physically within the brain, then certain types of memory need not necessarily be confined to individual minds, and Sheldrake interprets Jung's notion of an inherited collective unconscious as a kind of collective memory.[3]

Here we have the mutual reinforcement of two sympathetic ideas, that of a collective unconscious mind, and of a non-energetic archetypal form which both emerges from, and contains it.

Lyall Watson adds yet another angle to these concepts when he describes what he calls a 'contingent factor' in biological evolution.

> The contingent system is Jung's collective unconscious extended to include biological effects and factors, and given the possibility of physical location connected in some way with the symbiotic nature of all nucleated cells.[4]

Watson illustrates his meaning with the now famous story of Imo, the Japanese Macaca Monkey. These monkeys had been studied intensely in their wild state over a thirty-year period. Then, in 1952, man supplied what Watson calls 'the right evolutionary nudge'[5] by providing the monkeys with a new kind of food in the form of unwashed sweet potatoes.

Accustomed to a diet of buds, fruits, leaves, shoots and bark, nothing in the monkeys' evolutionary experience had equipped them to deal with the sand and grit adhering to the new food.

> Then an eighteen month old female, a sort of monkey genius called Imo, solved the problem by carrying the potatoes down to a stream and washing them before feeding. In monkey terms this is a cultural revolution comparable almost to the invention of the wheel.[6]

In due course Imo introduced further new behaviour patterns to the troop, but the most interesting thing of all was the way in which other monkeys learned to follow her example. At first it was those in immediate contact with Imo, who began to copy her actions. Gradually more and more monkeys on the island were washing sweet potatoes prior to eating them, until, with the addition of only one more schooled monkey, a critical threshold was reached. Beyond this threshold the newly acquired habit was enabled to jump the natural barriers of distance and space, and to appear spontaneously in far-away colonies on other islands, where learning by imitation was impossible. So far as could be judged from Japan's Macaca monkeys, the new capacity had become species-specific.

Watson uses the number 99 at random to illustrate the critical mass at which the addition of one more potato-washing monkey might precipitate a new capacity throughout the species.

Watson calls this emergent capacity the 'Hundredth Monkey Phenomenon', a mechanism in evolution other than those governed by ordinary natural selection, and suggests that it might account for the way in which many themes, ideas and fashions spread through our culture.

It is easy to see that Sheldrake's morphogenetic fields could be just such a mechanism, acting across both time and space as he describes. Their cumulative effect would be the archetypal species-specific norms described by Jung. Previously existing cultural patterns would be 'present' to subsequent cultures through the influence of morphogenetic fields.

The psycho-analytical schools postulate that many aspects of subjective experience depend on the sub-conscious or unconscious mind. Jung takes this hypothesis a step further, tying it firmly into the evolutionary experience of man by suggesting a common substratum of consciousness, shared by all human minds:

> In addition to our immediate consciousness, which is of a thoroughly personal nature, and which we believe to be 'the only empirical psyche, ... there exists a second psychic system of a collective, universal and

impersonal nature which is identical in all individuals. This collective unconscious does not develop individually but is inherited.[7]

When I first encountered Jung's work I was amazed and comforted, for I had come independently to a similar conclusion and was embarrassed at having come up with so bizarre a notion. Here was another nut, and it made me feel less lonely.

Just how the collective unconscious could be inherited is still a matter of debate, but it seems to me that if there is any hope at all for the human species, it must lie in the gradual impingement of the collective *un*conscious on the conscious paradigms of human thought. Individual attitudes and behaviour will thus become increasingly porous to the pressures of the collective unconscious which I have called the Gaian mind.

If Jung's idea is correct, then clearly Gaia already has an unconscious mind, and her point of emergence as a fully self-conscious entity will entail only the snapping into functional coherence of her pre-prepared cerebral equipment - the minds and brains of men. Her consciousness will be the 'emergent quality' arising from the interaction of a critical number of human and other life forms, but her final archetypal 'shape' will remain within the cosmic potential of the Earth cell.

Applied to the lives of men on Earth at the present time, these notions raise some rather frightening questions. If we are building archetypal species-specific norms which will be present to subsequent cultures through the influence of morphogenetic fields, dare we contemplate what their inheritance will be? Human behaviour patterns in our century must be creating morphogenetic fields too horrific to contemplate. But are there not perhaps Cosmic fields stronger than our own, reinforced by millions of years of experience elsewhere, which must be present to our planet even now (the over-riding impulse I've been talking about) to whose resonance we shall presently respond?

I can imagine the Gaian field resonating to this cosmic pulse, and we, mankind, responding in turn to her frequencies. We do not know what our critical threshold may be, but that there are forces at work beyond our own I have no doubt. The 'Hundredth Monkey Phenomenon' surely applies to us too. We must be very near to the count of ninety-nine.

CHAPTER THIRTEEN

4.6 The Time Factor

Summary: Evolution is speeding up.

When one thinks about the evolution of global consciousness, and adds to this already unfamiliar notion that one expects it to be well on its way to reality within the lifetime of people alive today, the colossal Time Scale of evolutionary process is an immediate and daunting stumbling block.

Supposing evolution to be, as I have suggested, analogous to the differentiation of tissue in the body of Earth, and looking back over the billions of years involved in the process so far, how is it possible for this colossal leap in consciousness to occur within the lifetime of a few generations of men? Is it possible? I believe that it is.

It is a well known axiom of evolutionary theory that environmental pressures are a prime stimulus to growth. Until very recently we were taught that change took place through millions of years in infinitesimal stages; so slowly indeed that the process was invisible during the lifetime of a man. Today there is a little healthy doubt about this. Evolution is seen to proceed in great leaps and bounds, and there is much discussion as to how this comes about. Whatever the merits of these conflicting views, it is clear that the reaction of an entity to its environment is a key factor in its development. We know that interface experience is a stimulus to growth. Complexity increases interface experience, and so must speed up growth. Complexity is thus one of the causal factors in the developmental process, and any increase in complexity should lead to acceleration in the evolutionary pace of the system.

This would seem to me to be true whatever the origins of life on our planet, and whatever the possible influence of additional genes from outer space, or of morphogenetic fields or of any other contingent factor that may exist.

So far as we know, nothing like human intelligence had appeared on Earth before a few million years ago. This, in the geological time-scale, is very recently indeed. Still more recent is the evolution of substantial intellectual abilities on Earth, which is a further indication that we are perhaps looking at a significant increase in the *tempo* of evolution, which, as I see it, is made possible by the increase in biological complexity over millions of years into ever more manifold forms. This increase in tempo is not, so far as I know, part of present evolutionary theory, but neither is it necessarily in conflict with it. Hoyle and Wickramasinghe argue very convincingly[1] that, contrary to Darwinian theory, the biggest evolutionary transitions must take place swiftly, a situation which they liken to a quantum transition in physics, but their conclusions are based on the mathematics of gene theory, and do not touch on the possibilities of acceleration due to complexity. They do however, lend substance to the thought that if such quantum leaps are indeed possible, we might be on the brink of just such a critical transformation as they describe.

Those same crucial genes that Hoyle and Wickramasinghe describe have enabled them, and people like them, to do their mind-boggling calculations, so anything is possible! Moreover there seems to be a great deal of latent intellectual potential to account for which is not explained by Darwinian theory. They quote from Dr Ohno's book *Evolution by Gene Duplication* as follows:

> Did the genome of our cave-dwelling predecessors contain a set or sets of genes which enable modern man to compose music of infinite complexity and write novels with profound meaning? One is compelled to give an affirmative answer ... It looks as though the early Homo was already provided with the intellectual potential, which was in great excess of what was needed to cope with the environment of his time.[2]

This non-Darwinian view fits in well with my own thought: The structures that might give rise to Gaian consciousness may already have been formed in the cerebral cortex of man, and perhaps only wait on some critical threshold factor to come fully and instantly into play, as in the Hundredth Monkey Phenomenon cited above.

Acceleration is the rate of change in velocity. It is measured as a change in velocity per unit time. Applied to evolutionary change it might have gone something like this:

As the Earth became more and more complex, the possible interface relationships between biological units would have become more and more numerous. Thus the stimulus to growth within the system would have increased from moment to moment through all those millions of years, and the whole process would have speeded up. The exponential nature of this acceleration will be clear to anyone familiar with the rudiments of mathematics (among whom, sadly, I do not belong).

Just as cerebral activity of the brain appears to depend in large measure on the manifold interconnections and micro-circuits made possible by the synapses between its millions of neurons, I suspect that the evolution of any system must accelerate in mathematical relation to the number and complexity of its internal units, and that this principle will apply in particular to cerebral aspects of the Earth organism.

This is another way of saying that the rate of acceleration in the evolution of the Gaian consciousness will be partly dependent upon the number of relationships made possible between men. If you add to these all the other conscious living forms from plant to animal, and try to picture the multiplicity of possible interface relationships between them, it becomes clear that the rate of acceleration in the evolution of the Gaian consciousness will be as near infinity as makes no difference.

Such an acceleration might bring a cosmic micro-organism such as the Earth into focus in relation to the cosmic Time Scale in a relatively short period. I think I mean by this that the Earth is perhaps developing as an integral living cell, at a rate consistent with its role in the cosmic body corporate, just as do the cells of our own bodies, and that the life-span of the Earth is perhaps comparable in the cosmic time scale to that of a single cell in the life of a human being.

When one looks at evolution as differentiation of tissue in the body of Earth, and when one thinks of men as possible neurons in its 'brain' it is clear that our focus of interest at this time must be the imminent evolution of consciousness in the infant global mind.

Peter Russell points out in *The Awakening Earth*[3] that the population of the Earth is expected to stabilize at about ten billion during the course of the twenty first century, and that this number bears a strange resemblance to the number of neurons in the human brain. In addition it appears to be critical in a number of physical and biological processes before major evolutionary change can take place. To make the quantum leap to Gaian consciousness, therefore, would seem to require only that all these human 'neurons' be brought into functional neuronal relationship. We have to establish the circuits and symbiotic relationships that will do this. I shall argue that for the first time in human history we have the tools which will make this possible.

It has often been said that human beings had reached a stage where we were able to manipulate the further course of evolution. I think that 'manipulation' may be too large a word for this. Seen as developing neurons in a global brain we could not be other than significant. We can surely create the future to a very great degree. But with luck, (since we are not always wise), the life-impetus of the planet itself will co-ordinate and contain its parts, and we shall gradually become aware that we are working within a keener global intelligence, following unconsciously those rhythms consistent with its overall field.

If we take a long backward view of the Earth, we see the differentiation of tissue into the myriad forms that so delight us now, interrelated, ever more enmeshed, ever more sentient. This for today. Looking further still, and forward, we might glimpse completion of the cycle - all these forms in functional cohesion - a breathtaking wonder for tomorrow.

BOOK II

The Application of Living Principles

CHAPTER FOURTEEN

5.1 Living Systems are Information Systems

Summary: The exchange of information is so basic to the organization of living things that it extends right through from subatomic particles to human brains.

The key to any solution of our global problems will therefore be information-processing.

Sooner or later you have to translate conceptual thinking into practical policy, and assess its usefulness in everyday life. I didn't like this part much at first. Coming to grips with the screaming nightmare of today's world isn't fun. But if we are to devise economic and political processes compatible with a living Earth, we need to look very closely at organic principles to discover which of nature's methods, if any, are applicable and relevant to the human situation. If we find that consistent criteria appear to guide the morphogenesis and operation of living organisms at every level of life, it would be logical to apply these same principles to human affairs, adding only those criteria of mind and spirit peculiar to human concerns.

We have already identified four essential systemic components in the behaviour of particles in the cybernetic activities of protein molecules.

They are all, I think, intensely significant, and worthy of the closest possible study in the human context.

They are these:

- All activities that contribute to the growth of a system are inter-connected and inter-controlled, directly or otherwise.

- They are self-governing.

- They depend upon free choice in response to information received.

- The criteria of choice would appear always to be their contribution to the coherence of the system.

These microscopic systems are stunningly competent, but they refer so far only to the intracellular activity of small single units. With the development of multicellular organisms, further considerations of structure and physiology arise which could conceivably be applied to human problems. I hope to explore some of them at a later stage in this book. For the moment I want to concentrate on a single fundamental question:

What, in any system, is the key to self-regulation of the kind cited above?

The answer lies, I believe, in information.

Whatever else they may be, and they are many things, living systems, from molecule to man, are information-processing systems. If Gaia is a living organism, we are going to have to look at her in similar terms. The key concept in any global solution to our complex human problems will be information-processing. I think this is where the computer comes into its own, but more of that later. If living creatures are essentially the self-governing cybernetic entities they seem to be, it will be in their methods of information-processing that we are likely to find clues most fruitful to our needs. It looks as though the prime factor in our global approach to sanity is going to be an exchange of information as pervasive and unfettered as that in other living systems, and though the methods used by nature are by no means fully explored, they are neither new nor untried. Life has been sustained by them for more than three billion years.

How, then, do living systems communicate?

To answer that we have to ask an even more basic question:

Is there a means of information-exchange so fundamental that it operates at every level throughout the physical universe? And of course there is. We've met it before in some of the concepts of subatomic physics already mentioned. Every part of the universe 'knows' what is going on elsewhere and is affected by it, because 'at the level of implicate order every part of the universe contains the whole universe enfolded within it.'[1]

We remember that (p.8) 'when two particles have interacted and then flown off in opposite directions, interference with one particle will instantly affect the other particle, regardless of the distance between them.[2] Somehow information has passed between the two. This sub-atomic 'knowing' must surely be the ultimate in information-exchange!

Moving on to molecular structure, (Ch.9) we remember that a protein molecule is capable of 'regulating its activity in response to information emanating from several chemical sources,[3] and of 'choosing to activate a reaction' in response to such information. Clearly information-exchange is very basic indeed, as is individual choice, not only at the sub-atomic level, but in the formation of that most crucial element of living tissue, the protein molecule.

Most of the key processes in our bodies depend on protein molecules. The largest of them comprise tens of thousands of individual atoms self-organized into sub-units, which in turn join together to form the more stable molecule, all of which apparently effortless self-organization takes place according to the laws of information-exchange applicable to living chemical systems.

What fascinates and especially delights my questing mind with global synthesis at stake is that this microscopic information-processing leads to the *spontaneous self-assembly* of molecules:

'It is relatively easy to break down a protein into its sub-units,[4] observes Jack Weber, 'but even simpler to build it up again. Shake the pieces together and the three-dimensional jigsaw will spontaneously reassemble into a perfectly formed protein. The process requires no special treatment and it will even happen in a jumbled mixture of many different proteins.'

This astonishing phenomenon suggests that the spontaneous self-assembly of parts into wholes in response to information is a basic characteristic of life.

Weber remarks that the fact that very complex molecular structures are able to assemble themselves in this way has been an important finding for scientists who are trying to understand how the first simple life forms came together. It may prove equally significant in providing clues to the self-organizing structure of larger units, in particular those relating to the social structures of man, for its complementary qualities of free association and freedom to choose are very dear to his heart.

We have already seen (Ch.4) that the work of Prigogine and others on dissipative structures is being applied increasingly in related disciplines, that bio-chemical and biological processes are widely viewed as extensions of the principles found in dissipative structures, and that this study has been applied to the study of many social groupings in the animal kingdom. Finally it has been found to have relevance 'to human economic interactions, to human society, to ecosystems and even to Gaia herself.'[5]

Fritjof Capra develops the systems view further in his splendid book *The Turning Point*, published in the same year (1982) as Russell's *Awakening Earth* just quoted above.

> Systems theory looks at the world in terms of the inter-relatedness and inter-dependence of all phenomena, and in this framework an integrated whole whose properties cannot be reduced to those of its parts is called a system. Living organisms, societies and ecosystems are all systems...

Capra goes on to explain that living systems are organized in such a way that they form multi-levelled structures, each level consisting of sub-systems which are wholes in regard to their parts, and parts with respect to larger wholes.

> In fact we shall see that parts and wholes in an absolute sense do not exist at all.[6]

Capra quotes Arthur Koestler as having coined the word 'holons' for these sub-systems which are both wholes and parts. Each holon has two opposite tendencies:

> ... an integrative tendency to function as part of the larger whole, and a self-assertive tendency to preserve its individual autonomy. In a biological or social system each holon must assert its individuality in order to maintain the system's stratified order, but it must also submit to the demands of the whole in order to make the system viable ... These two tendencies are opposite but complementary. In a healthy system - an individual, a society, or an ecosystem - there is a balance between integration and self-assertion. This balance is not static, but consists of a dynamic interplay between the two complementary tendencies, which makes the whole system flexible and open to change.[7]

Our current political, judicial and social systems are all unsuccessful attempts to balance these two opposite but complementary tendencies. One does not need to look far to realize they are not working, perhaps because they were based on the mechanistic concepts of nineteenth century physics, whose rational, linear thinking patterns have little in common with the organic world. Living systems simply do not work this way. Our preoccupation with hard technology, our wasteful consumption, our ruthless exploitation of natural resources and our obsessional concern with growth, are all totally out of line with the complex interplay of non-linear relationships which characterize living systems.

The problem is compounded by the singular nature of the overall goal. At first sight nature appears to have moved from single cell to multicellular organisms in a straight progression from simple to complex. But the protein molecule was already enormously complex at the time of formation of the single cell. So the progression, in so far as we are able to observe it, is from atom to complex molecule to complex single cell, then on to the multicellular hyper-complexity of the articulated organism, and finally back to the unitary 'super-cell'again, which gives the process a cyclic or rhythmic nature rather than a linear one. To what larger cosmic order our global 'super-cell'may belong we cannot begin to know, but we do know we've got to make *this* cell work. It isn't proving easy, but I think it can be done, now that we are beginning to understand the nature of the system to which we belong.

Within this cell I see the possibility of a state of man in which information replaces other means of exchange in fulfilling human needs, just as it does in all living systems. I foresee a time when man's activity is pursued for satisfaction or usefulness or sustenance; for educational enrichment or beauty; for kindliness or spirituality, but whichever way, self-chosen. I see a world in which satisfaction rests in what one is and does, rather than on what one has or holds; in which what one has shall be useful, beautiful and appropriate; in which no man willingly hurts another for the simple reason that he does not feel separate from him; and in which each man judges his own actions in the light of the overall Gaian impulse to which he is tuned, in the deep understanding that unless Gaia is similarly attuned to a holy and living universe, no good can come. In the movement of each man's spirit, Gaia will move toward her God.

To accomplish this transition without self-destruction demands a degree of humility. We may have to borrow widely from the extraordinary diversity of lower life-forms found in nature, adapting and extending their techniques to accommodate human social and economic functions. Transportation and circulation of essential foodstuffs or minerals, the symbiotic use of waste products, and selective usage of available sources of energy are all areas in which we could benefit from close study of natural processes. But our central problem is that of integrating the thought of millions of self-oriented, very determined individual units, each convinced that his interests are in conflict with those of his neighbour. And this one we are going to have to think up for ourselves, for no one, so far as we know, has done it on quite the same scale before. Grab-what-you-can is still the rule among men, as it must have been in nature in the days of the single cell.

Evolution solved this dilemma with respect to the formation of multicellular organisms in a new and unique way by developing cells specialize for cybernetic processing, a development that was to reach a peak in the human brain.

In the next section we will take a look at this process with a view to its application in human affairs.

CHAPTER FIFTEEN

5.2 The Amazing Brain

Summary: This chapter is concerned with the automation of basic physiological functions such as breathing, eating, heartrate, or chemical balance, which occurred in the brains of higher mammalian forms prior to the development of advanced learning and memory.

We suggest that a similar automation (made possible by computers) may be required for the regulation of fundamental social and economic organization before true global consciousness can be manifest.

During the half-billion years that followed the appearance of the more sophisticated cells (those which would include a nucleus and mitochondria), cells first bonded together in multicellular organisms such as the sponge, in which information exchange was perhaps limited to an awareness of temperature, of currents, or of nutrients dissolved in the water which nourished them. Sponges had no nervous system with which to initiate a positive behaviour sequence, and were therefore entirely at the mercy of their environment. Then, with a fanfare of trumpets (or so it should have been!), life produced the first specialized nerve cells, such as we find in the jellyfish, and information-processing was on its way.

'So far as we can determine,' say Robert Ornstein and Richard F. Thompson in their book *The Amazing Brain* (1985),[1] 'nerve cells in every animal from jellyfish to human have the same basic electro-chemical mechanisms for conducting information. It appears that the primitive mechanism of the jellyfish nerve-cell worked so well that it became fixed in evolution. In order to generate more complicated and adaptive behaviour, all you need to do is put more of these nerve cells together in more complicated ways.'

The ways in which nature performed this feat produced, in the course of evolution, millions of different species, each of whose complicated ways of assembling nerve cells was peculiarly adapted to the environmental conditions in which it had to survive and the tasks which it had to perform.

The human brain assembled ten billion such cells, and since control of bodily processes is a small part only of human brain function, their central activity concerns the extraction of patterns and meaning from the

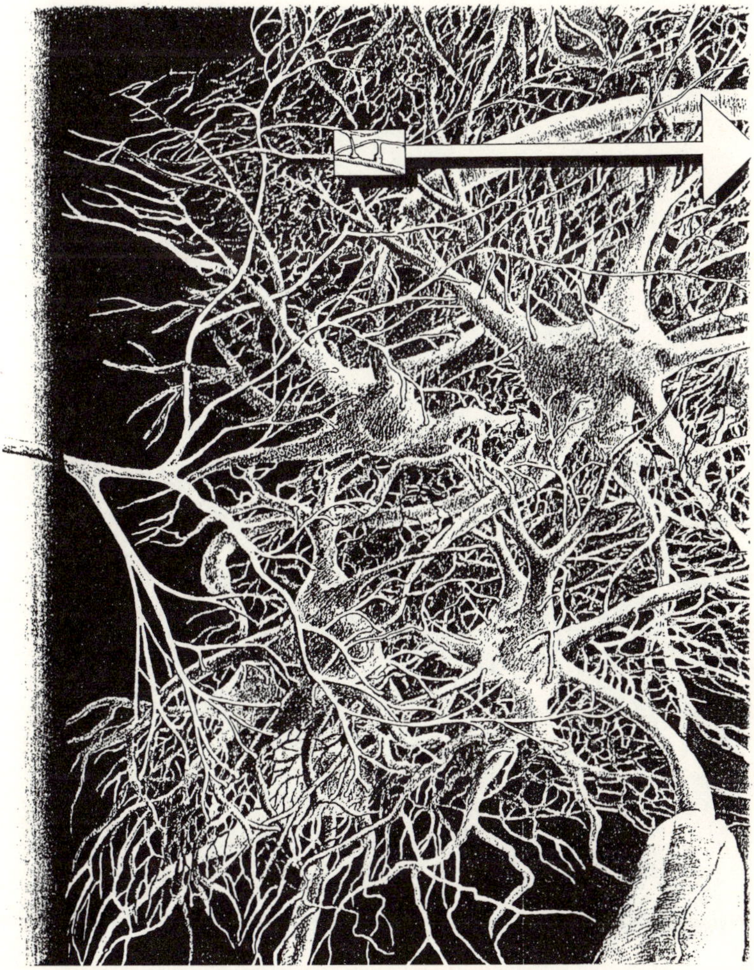

Illustration from *The Amazing Brain* by Robert Ornstein and Richard F. Thompson, Chatto and Windus, 1985.

Illustrated by David Macaulay.

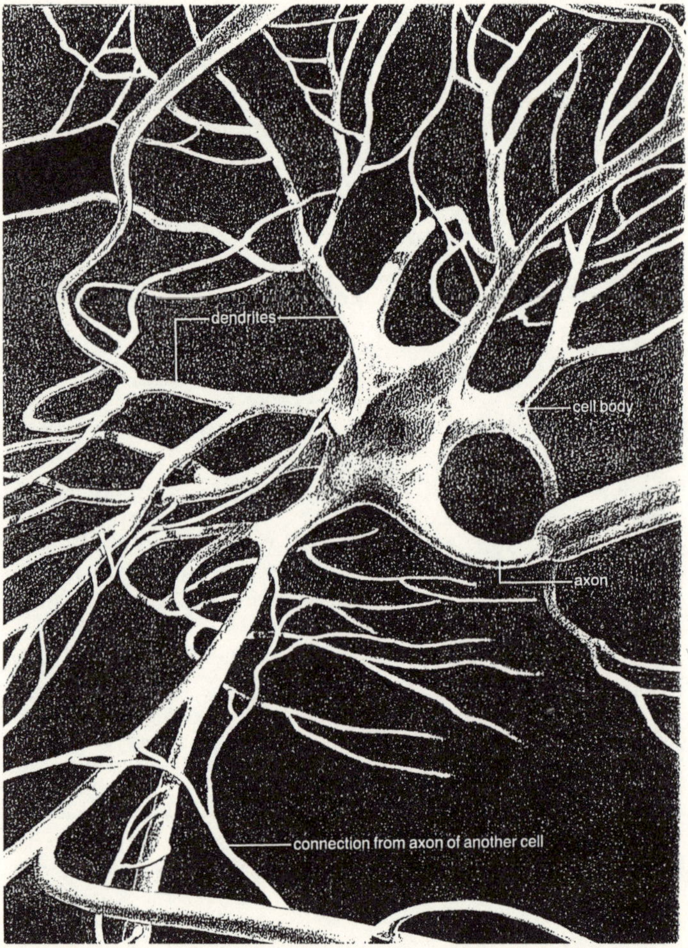

Illustration from *The Amazing Brain* by Robert Ornstein and Richard F. Thompson, Chatto and Windus, 1985.

Illustrated by David Macaulay.

mass of information bombarding them.

Each nerve cell in the brain processes ceaselessly the information coming in from thousands of other nerve cells and from chemical messages in the blood, and is in constant communication with them, receiving, processing and sending out signals whose cumulative effect is to generate behaviour and mood, as well as to control the vital bodily functions dependent on them.

Control of these vital bodily functions by the central nervous system follows an orderly and easily recognizable pattern in all vertebrate animals, and can be seen in even the most primitive animals that are bilaterally symmetrical, such as worms. These animals all have a left and a right side. The bodies of all such animals are segmented. Each segment contains bundles of nerve fibres that send information from receptor cells in the skin to a group of nerve cells that in turn projects information back to control the muscles.

The purpose of all this to-ing and fro-ing is co-ordination for purposes of movement, nutrition, circulation, breathing and the elimination of wastes. Compared with the complexities of 'higher' cerebral activity it is all relatively straightforward, and a great deal is known about it, so we might just be able to find some clues here to the broad regulation of global life-support systems as they apply to human beings.

The human brain is of course immeasurably more complex, and suggests itself immediately as the obvious model for global integration, but in spite of the enormous amount of research on the brain in recent years, its mysteries seem to deepen the more we know. However, from what we do know, a few broad principles seem to me to have relevance in our search for solutions to immediate human problems.

The human brain seems to have been built in three broadly recognizable evolutionary stages: first the hindbrain, the oldest part of the brain, which evolved more than five hundred million years ago. The hindbrain includes most of the brainstem (primarily concerned with the control of breathing and heart rate) and the cerebellum, attached to the brainstem, which developed initially as a motor structure to improve control of balance, body position, and movement in space, but now seems to have taken on other functions as well.

> It now appears that memory for certain types of simple learned responses may be stored there, particularly in those parts of the cerebellum that evolved most recently.[2]

Ornstein and Thompson observe that the assignment of new responsibilities to the cerebellum is typical of the way the brain has evolved.

Old structures were not discarded but rather expanded to handle other functions. We need to remember this in our efforts to design new structures for the global co-ordination of human society.

In the centre of the brain is a group of cell structures evolved between three hundred million and two hundred million years ago known as the limbic system. This too seems to have assumed different functions with the passage of time from those for which it was developed, thus reinforcing the realization that old structures can, and in living creatures do retain their usefulness by adaptation to a new role. The limbic system (or mammalian - so called because it is most highly developed in mammals) is the area of the brain that helps to maintain homeostasis, or a constant environment in the body. This it does by a process of feedback controlled by the incredible hypothalamus, no larger than a pea, a tiny, intricate group of cells that regulates eating, drinking, sleeping, waking, body temperature, chemical balances, heart rate, hormones, sex and emotions, and also by control of the pituitary gland which regulates the production of hormones. The operation of feedback control by the hypothalamus is perhaps the prime example of information-processing in the whole saga of bodily function, and might well be found to give us further vital clues toward the synthesis we seek.

I am concentrating for the moment on those aspects of information-processing in living systems concerned with the automatic regulation of basic physiological functions, because these seem to me to correspond at a global level to the fundamental elements of social and economic organization we have failed in so dismally so far. If nature follows the same sequence in the synthesis of global mind as she seems to have done in the evolution of homo-sapiens, then these survival processes will have to reach a point of automatic self-regulation before a significant burgeoning of consciousness becomes possible.

I think we are at a point, in this decade, where our technology is sufficiently advanced to give us the tools we need for this kind of global automation, and that we only need a change of direction in human thought to give us the vision to use them.

That word 'only' is perhaps deceptive. Changing the paradigms of human thought is a major challenge, as many have found to their cost. We don't like being jolted out of accepted patterns of belief. But the notion of a living Earth is more than a notion now. It is a formative concept whose time has come - perhaps the only concept whose practical applications can save us from ourselves. For the author it is even more than this. It is fact. Gaia not only functions *like* a living system. She is one, as much a living entity as you or I. The chief difference between us is one of scale, and that Gaia outdoes us in majesty.

The practical applications of concept that follow in this book are based firmly on the nature and capacities of living organisms, applicable to our predicament specifically and only because Gaia is considered to be one of them. Her tissues embrace all living creatures on earth, including ourselves. All are interlinked and interdependent; some are motile, others relatively fixed, as in bone, or the trunk of a tree. The parts and the whole have evolved together and continue to do so, flesh of one flesh.

We can neither trace all the pathways and connections of so vast a network, nor predict just where her intricate self-organization will lead. But if, in our dealings with one another, we follow as best we may the observable methods of other self-organizing living systems, we cannot go far wrong.

CHAPTER SIXTEEN

5.3 Aspects of Self-Organization

Summary: Can we achieve automation of basic social functions without loss of personal identity or initiative?

The particular aspects of self-organization I want to explore in this book are:

- Can we use the new technologies to assist us in automation of basic social and economic processes in such a way that they correspond to the automated processes of other living systems?

- Complementary and essential to the first, how can this automation be achieved without prejudice to the cherished autonomy of each human being?

The first question covers a lot of ground, and will form the substance of most of the latter part of this book, so I want to look briefly at the second one first:

Selfhood and the maintenance of identity.

How can synthesis and automation be achieved on a global scale without prejudice to human freedom?

The answer to this latter question stems partly from the nature of man himself, and partly from the nature of symbiotic processes, which work precisely and only because of the *differences* between their constituent parts. The way in which these differences complement each other is the basis of symbiosis. We'll be coming back to this.

We saw in the section on cells, as well as that on morphogenesis, just how persistent the preservation of selfhood can be, even at the lowliest levels of life. But in man this self-awareness has become so intense, and our insistence on human 'rights' so unyielding, that it has completely overshadowed the need for synthesis.

At the same time there has been an interesting shift in concept in recent years as to what those rights should be. As long ago as 1969 Richard Farson used this shift in emphasis to inject a note of optimism into Doomsday thinking. In spite of a lively realization that we perhaps faced the prospect of irreversible pollution and destruction of our natural environment, of mass starvation in the southern hemisphere as the industrialized northern half grows richer, of the possibility of nuclear war, and of a tumult of violent conflict affecting the entire globe, Farson insisted that the present crisis in human values sprang from an entirely new view of human potential. In this view, people would begin to demand above all the right to fulfil their own potential. This fulfilment, and not material possessions as such, would be seen as the means to a good life.

> For the High School Seniors of 1984, the good life will be focused on *experiencing their human-ness* ... and the purpose of life will not be to use themselves for ulterior goals, but to experience themselves; not to use others, but to experience others; not to use their environment, but to experience it in the fullness of its possibilities for richness and beauty.[1]

The year 1984 has come and gone and the confusion persists, but insistence on human rights remains perhaps the strongest single motivation of a variety of diverse groups throughout the world. We band together to insist on our right to be apart! And we shall find that this is not exclusively a human trait. We have already seen it at work in the intra-cellular life of organelles Lewis Thomas discusses so charmingly. But it is equally evident in the sensory equipment of articulated multicellular organisms, and in the neuronal arrangements by which sensory data is conveyed to the brain for processing, as for instance from the human eye.

These visual arrangements consist of a network of nerve cells strung together for the express purpose of conveying information from eye to brain, yet each cell preserves its essential autonomy. Hansjochem Autrum explains how this is done in his essay on 'The Communication Network of the Human Body' (*Man and Animal*, 1972).[2]

The visual cell produces a weak electric impulse when light falls upon it. But, says Autrum, this potential is not transmitted directly through the nerve fibres to the brain, the way we conduct the potential of our electric generating stations.

> Nerve fibres have a resistance *a million times higher than that of electric cables*; so that there is no possibility of a direct transfer of potential.[3]

Instead it is the nerve fibres themselves which produce electric potentials at every point. Autrum comments that they resemble a long and extremely fine chain of single electric batteries.

This miracle is wrapped in a covering not much thicker than a millionth of a centimeter, composed of a superb insulating material. Insulation is especially necessary because the covering, or membrane, is maintained at a potential of about one-tenth of a volt by the nerve fibres. One tenth of a volt is very high for a battery of such minute dimensions; if we translate this into terms which we can understand, we find that this means that over a distance of only one centimeter there is a potential of *one hundred thousand volts*.[4] (The italics are mine).

In the context of individual identity the astonishing insulation of nerve fibres is a very interesting phenomenon. There is a sense in which it illustrates the extreme of individualism. It suggests a kind of personal privacy at the cellular level, ensuring maximum personal integrity even under conditions of functional co-operation as close as those to be found in the human nervous system.

As an analogy for co-operation between people in a living planetary entity, such sturdy selfhood is very heartening. Each unit operates on its own immediacy, insisting on its own intrinsic value, its own potential. Yet each engenders energetic action in the functional competence of the whole. We need to remember this in any attempt to apply natural principles to the automation of human social patterns. It could hardly be more important.

Symbiosis

We come now to the all-important question of symbiosis. We have looked briefly at some of the principles that seem to emerge in a study of the internal structure of multicellular organisms. I now want to examine some of the relationships that occur in nature between these multiple systems in relation to each other, while falling short of total organic fusion. They range from the nitrogen-fixing bacteria in the roots of leguminous plants to the well known colonies of bees and ants, whose patterns of co-ordination and collective intelligence far surpass anything that any of these creatures could achieve on their own.

Bacteria in the tissues of animals usually live in mutually beneficial (symbiotic) association with their hosts, increasing nutritional performance and increasing resistance to disease. There are birds that live in close association with grazing herbivores, controlling parasites or eating insects the animals disturb. Fish clean the teeth of larger fish in their pursuit of food. Medusae become entangled with snails in a relationship so bizarre one can no longer say which is which - only that neither can exist without the other. The seeds of certain trees need to be digested by birds before they can germinate. The list is endless. Many species and types of creature exist in such intimate biological association they appear to be one species. The range covers micro-organisms, plants and animals in every imaginable combination. All involve the symbiotic principle on which the whole of nature depends, so much so indeed, that symbiosis cannot but be regarded as a pre-requisite for ordered life.

On the borderline between mutually beneficial free association and total fusion, is one of the strangest encounters between primitive entities in the whole of nature. This is an encounter between a colourless fungus and a green alga, whose partnership reduced the ancient rocks and gave us soil - a very early example of reciprocal arrangements between cells. In a world whose complexity boggles the mind, it has the added advantage that the participants, at least on the surface, are limited to two! It is the lichen story.

One finds in this primitive life-form, the lichen, those surprising qualities we have come to associate with molecular and cellular development all along the line: selectivity; mutually beneficial association; preservation of identity - with the added bonus of beneficent global repercussions way beyond the dynamic of each tiny life. That these principles will have relevance to our specifically human situation will, I hope, become clear as we go along, and may even throw light on problems that presently seem intractable.

I know of no other account of the lichen half as fascinating as that in Rutherford Platt's most beautiful book *The Great American Forest.*[5] I am loth to fragment Mr Platt's work, but quote perforce only those scraps which have direct bearing on my own work. Each word sparks a tiny flame in the mind.

The partnership of fungi and algae arises from reciprocal need, a common arrangement in nature. But the fungus-alga-combine is unique in the way it joins two different kinds of plants *to create a third kind*. Fungi provide the fabric of the lichen, and trap the necessary moisture. Algae perform the service of photosynthesis by which the partnership is nourished.

The bodies of fungi are colourless tubes - exceedingly fine threads which grow very fast, snaking in all directions to form extensive, tightly knit, leathery networks. They have no chlorophyl.

The algae that merge with these fungi are microscopic, spherical, one-celled organisms. They exude a gelatin by which they cling together, end to end, in short chains.

In the lichen, the green cells of the alga occupy a particular spot just beneath the upper surface of the leathery fungal body. Since the fungus threads hold water, and are also translucent to light, they provide a tailor-made habitat for the algae, which require both water and light for the process of photosynthesis.

The fungi, on the other hand, find themselves equipped with a food-making organ, because algae use sunlight and carbon-dioxide to make sugars, after the manner of green leaves.

'A lichen', says Rutherford Platt, 'is a highly successful combination, with more power for survival than any other many-celled living thing on earth.'[6]

> It is an air plant in the sense that it takes no nourishment from any living or dead material to which it clings. A lichen on a tree trunk clings to the bark with holdfasts that do not penetrate living cells or hurt the tree in any way. This is the only kind of plant that can flourish in an utterly sterile place. For this reason the lichen played a heroic role in the spread of the forest northward across ice-scanned rocks in the wake of the ice-age glaciers.[7]

Platt describes how a lichen attaches itself to the bare face of a rock by brewing a powerful acid which dissolves rock crystals, and then pokes a tough thread, or holdfast, into the tiny cavity between grains of lime in limestone, and between crystals of mica and garnet in granite. The salt crystals in the lichen reflect the spectrum of sunlight 'like minute diamonds.'

> Even silicate, granite's hardest crystal, is dissolved in time by lichen acid. In this way, the crystals of granite are eventually loosened. A speck of sand results. A bit of organic duff is made. The weathering of rock is promoted and with passing centuries a place is prepared for trees and flowers.[8]

One would expect that so close an association as exists in the lichen between algae and fungi would result in a loss of identity. A new species has actually been created. But not a bit of it. As in every other natural process we've explored so far, the identity of both fungus and alga is wholly maintained throughout the partnership. If they are separated, and properly nourished, they simply pursue their separate lives.

We are moving gradually away from the idea that because the Earth is so complex, and man so rugged an individualist, the basic laws of life can no longer apply to him. *All* natural systems are complex, even the lichen, if you consider the make-up of each single cell. Each survives through interaction with its external environment (including its fellows!) and by organizing its internal components in such a manner that it is simultaneously an appropriate environment for them and itself a coherent, functioning whole.

All the components at every level are individuals. Internal and external environmental factors impinge on each other by way of information in the form of energy, and this is received and exchanged in a multitude of ways. All are inextricably enmeshed with all, through a complex system of information exchange. This is the key. The system selects from the flood of chemical and electro-magnetic information impinging upon it only such substances and frequencies as are appropriate to its needs. Within each 'simple' cell is a universe.

In nature what makes the whole thing workable is its diversity: diversity of habitat; diversity of species; diversity of behaviour patterns and need; diversity of satisfactions. Without this enmeshment of diversity, the interlacing reciprocal effects would be missing, and the whole thing would fall apart. We need to encourage these supportive effects, conserve them, go along with them, express them fully in human affairs, even when their benefits are not immediately observable. It follows that we must not try to reduce people to one common denominator - 'reduce' is the operative word! The ideal is not identity of type, interest or possessions, complete to the last robot or microwave. It is the fulfillment of men, and the richer the variety of type, loves, capacities, interests and satisfactions involved, the better the thing will work.

CHAPTER SEVENTEEN

5.4 Development of the Embryonic Global Brain

Summary: A number of human species-specific social and cultural structures may have to be developed to provide the neuronal links required, before global cerebal coherence can snap into function.

When you compare the political order with the simplest of nature's self-governing systems, it is impossible not to regard our attempts at government as pathetically crude. We have to find our way to something new, something organic and therefore self-regulated, something to engage the highest potential in men. But this is a task so dauntingly complex one would give up without even trying, except for one thing: Hard though it may be to believe, in the genetic history of virtually every multicellular structure known to biology, each of thousands or millions of cells 'carries within itself the whole of the species' system of responses, obeys and commands at the same time; each and every cell is simultaneously master builder and universal workman.'[1]

Otto Koehler describes this phenomenon in his essay *Chance, Necessity and Plan in the Living Universe,* with reference to the very humble life-form *Dictostelium* which grows on the dung of grazing animals.

We've met *Dictostelium* before as a forest slime, and learned how the multicellular fruiting body bursts open to release an individual single-celled amoeba again. Koehler describes how, in the laboratory, an amoeba will creep out from each spore onto a plate of agar which has been inoculated with bacteria; the amoebas swarm, eat and procreate.

Suddenly the confusion ceases. The amoebas creep together to form columns facing in one direction; the rear-guard then presses to the front and forms a tightly packed wedge at the head of each column, the amoebsa now standing rigid one above the other about the axis of the column to form stem cells. Other amoebas crawl up the outside of each stem till they reach the top, there becoming spores.

So here we have thousands of single-celled organisms suddenly combining in a well co-ordinated manner in accord with the structural plan of the species, forming the multicellular sporophore.

It has been proved to be a matter of pure chance where, on the agar plate in the laboratory, the centres of activity will form, at what point the columns will unite, which amoeba will end up in which fruit body, which one will become stem and which will end up a spore.[2]

We're getting used to the idea of indeterminacy in natural law, though in the formation of higher multicellular organisms I incline more to choice than to chance. If there is 'choice' in the structural activity of atoms in complex protein molecules, how much more might there not be at the cellular level described above? The crucial point is this:

If, as I believe, the Earth is a developing, embryonic organism, in the process of becoming self-conscious, and if I, and everyone else with me, are cells in its multiple body or brain, then I, and everyone else, must carry within me in latent form, the whole of the Earth's potential system of responses. We ought to know instinctively what course to take to further coherence on Earth. Why don't we then? How is it possible for human society to find itself in the sorry state prevailing today?

We forget, perhaps, how very young we are. We are in a stage of turbulent, over-eager growth, multiplying, spreading over the earth. We get in each other's way; fall over each other; struggle to stay on top. But we are learning. We have to - to stay alive. We are no different in this from other new-born things. Instinctive genetic traits take time and usage for development. We have to pinpoint the factors which trigger them.

Koehler discusses this limitation in the same essay quoted above. He suggests that the human baby arrives in the world like a kitten, entirely ruled by instinct. His mind develops gradually at first. Once he starts learning his mother tongue the infant matures in a different way.

> If it were not innate in him to breathe, cough, and sneeze, if he were not born with the ability to suck, swallow, and digest, he would have suffocated or starved to death long before he had time to learn all these actions.[3]

Koehler emphasizes his point further by pointing out that the same phenomenon is observable in birds.

> As soon as the feathers of the young swift are fully grown, it throws itself out of the nest into the air, avoids every telegraph wire, and returns safe and sound. If birds had to learn to fly from scratch, every one of them would land on the ground, shattered at the first attempt.[4]

Koehler makes the point here that those species-specific motor patterns which cannot be performed straight after birth are not learned.

The organism has to wait until the necessary structures are fully

> developed; *then the entire capacity abruptly becomes available* and learning can be added only by applying it.[5] (The italics are mine)

If we look at evolution in this light, then each evolutionary 'leap' - the moment at which the young swift hurls himself from the nest - had to wait for the necessary structures to be developed before it could come into being. Soil formation, atmospheric composition, plant and animal growth, all the fantastically complex interrelated features of what we call nature had to develop first, before the advent of man. In the same way a number of species-specific social and cultural structures may have to develop in man to provide the neuronal links required before global cerebral coherence can snap into function.

Since Gaia is a living organism, the growth of these structures will proceed on the impetus of her own coded energies, of which we are a part. We do not have to invent them, structure them or enforce them - only envisage them. But by stretching our minds, and tuning our being to work within, instead of against them, we shall help ensure the healthy growth of the embryo. Working against her could mean deformity or death.

So here am I, trying to resonate in tune with this living child - listening for God - trying to tune my creative urge to the source of things. In those moments of expanded consciousness, I get a glimpse of her.

Gaia - child of the universe - beautiful, bountiful, wholly without flaw. Down here, where we mostly are, there is turmoil and death. How do we reconcile these two?

Just as a child with a specific genetic background can, within the constraints of its environment, develop or fail to develop its best potential, so, I believe, can Gaia. And since we, in part, are Gaia, much of it will be up to us. I am not a futurist. I can neither make models nor attempt to predict the future. Nor do I want to. The near-infinity of variables in a living system make it totally unpredictable. Hence its interest and its wonderment. But I do think by using our creative imagination at fullest stretch, we can incline the trend this way or that, playing a useful role in a future we may not predict.

This, I think, is what I am trying to do in this book. It is rather like trying to choose the best route on a road map, with little or no idea of your destination. So now I'm going to try it the other way round, imagining a future of dynamic equilibrium, and then trying to plot, if I can, a route towards it. Because the potential is somehow coded in all of us, some of it will come true.

These are the Gaian faculties I foresee:

- *Complete intuitive coherence of all parts of the biosphere*

The biosphere is conceived as the blanket of living tissue enveloping the earth, roughly embraced by what we call 'nature'. It includes mankind. Its physical coherence is an accomplished fact. The Earth was fair and functional at an ecological level until man came. Here our function is not so much to imagine as to learn, and to restore.

- *Complete intuitive coherence of all parts of the noosphere*

The word 'noosphere'was coined by the Jesuit Priest, Pierre Teilhard de Chardin, to denote 'the cumulative effect of human minds over the entire planet', and was envisaged by him as a single interthinking group. We are crucial here.

- *Communion with God*

Coherence of the spiritual life of mankind concerns the relation of each individual man to his God, and can be pursued by him alone. I have nothing to say about it here, save that it is the fountain of life.

- *Enhanced consciousness of men*

This, I believe, will follow as the night the day, as Gaia moves towards self-consciousness.

- *Global self-consciousness*

For this Gaia has to wait until the necessary structures have been fully developed. Then the transition will be as abrupt as the bird.

- *Global communion with the universe*

A little beyond my scope, this one - but surely inevitable?

It is the second Gaian faculty, embodying the synergetic coherence of human thought with which we are chiefly concerned in this book. From it will spring those forms of human activity appropriate to a living Earth.

PART 6

Towards Synergetic Coherence in Human Activity

A Look at Current Thinking

CHAPTER EIGHTEEN

6.1 The Problem Defined

Summary: How can the transformation be made from a money-oriented, power-ridden consumer society to an information-based living organism?

At this point in the conceptual game I pursue so eagerly, one question is paramount. What is the task of ordinary men and women like myself sufficiently in love with life to want to give something back? How can we help? I'm trying hard to find out.

I've got to the point of feeling that our present socio/political/economic systems are dangerously inadequate. I conceive of their being replaced by a global information network such as we find in other living organisms. I am trying to find out how this could be done. As a first step, since computers and telecommunications are characteristic of our immediate world, along with economic crisis, threats of ecological disaster and widespread starvation, I'm exploring the possibility that the one could be used to help correct the others; that it could do this by automating the production and flow of material goods and services; and that in so doing it should emulate, and would parallel, the autonomic nervous system of human beings themselves. I suspect that such automation is a necessary prelude to a quantum leap in consciousness for the planet as a whole, and that the higher cerebral functions of mankind and of Gaia herself wait upon such automation.

The immediate problem, then, is how to make the transformation from a money-oriented, power-ridden consumer society to an information-based living organism. Put more simplistically yet, it means how to change the means of exchange from money to information.

I suspect that in the normal course of evolution, conditions have developed which have progressively rendered obsolete both the profit-oriented market economy and the centrally planned economy, as well as all the degrees of money-oriented combinations of the two presently in use, and that a comprehensive, self-regulating, global information network will do the job far more efficiently.

I am suggesting that we now have electronic tools to do this job, making it possible to ensure thereby, without coercion, an acceptable level of material wellbeing for every man, woman and child on Earth, while preserving nature's life-support systems in all their wonder and beauty.

I am suggesting that this utopian-sounding co-ordination of global supply is no more difficult nor impossible to achieve than in any other living organism, and that we can, by using our much-vaunted intelligence creatively, encourage and promote such a system deliberately, thereby avoiding some of its pitfalls and speeding up the process.

We have already taken the first step, in the sense that we have begun to *imagine* a new society. We have a new realization of the merit inherent in fluidity; of the possibilities of creative, imaginative change; of the infinite variety of pattern available for choice and synthesis. We have an imaginative capacity given to no other living animal (a capacity crucial to our survival), and we have the tools whereby to implement our vision. We begin to flex our muscles ready to use these tools. I hope to show in this section that our computers and telecommunication systems, properly programmed, can serve as a first step in the growth of a global nervous system, thereby regulating the flow of basic necessities throughout the planet.

One of our most intractable problems today is the disparity between the 'haves' and the 'have-nots' of the world, necessitating degrading and largely ineffective hand-outs from rich to poor, a surplus of food side by side with millions of hungry mouths, and the increasing apathy and helplessness of people slowly dying of malnutrition. You don't get this situation in a healthy multicellular organism. Either the whole organism is nourished or the whole organism dies, and this is how it must be with us.

We now have the tools. We can develop an information web so all-pervasive that it can regulate the flow of food, self-regulate local production, ensure a balance between activity and sustenance, and free the human race from greed and power.

What I am feeling towards is neither a market-regulated production-for-profit, capital-accumulation system, (whose extreme is the polarization of rich and poor and the concentration of capital in fewer and fewer hands), nor a planned production-for-use superimposed from above by a bureaucratic elite (whose extreme is the nightmare of a social machine in which men and women are no more than cogs), nor even a judicious mixture of the two.

I am looking instead for a spontaneous production and exchange of goods and services on the basis of computerized data concerning personal wish and need, in which computer models are all based on the value of the individual, and in which the wellbeing of the individual is perceived as being synonymous with the wellbeing of other individuals, as well as with the planet as a whole; in which money as a commodity has disappeared, and money as a means of exchange has been superseded by information; in which the renewable resources of the planet are conceived as its true capital, and over-exploitation of non-renewable resources has disappeared; in which the 'market' has been replaced by a network of information which self-regulates all production, whether local, regional or global; in which all information is universally available and power-wielding an anachronism; and in which the individual himself decides what he wants to do with his life.

I see computers, terminals, data banks and information-processing equipment of all kinds so widely distributed and interlinked that central control by a computer-aided bureaucracy is beyond possibility; I see socio-economic relations at once so personal and so fluid that their networks form and dissolve spontaneously and continuously as circumstance and inclination dictate. I see an economy that is at once sustainable, ecologically healthy, and satisfying to man: and I see the impetus for all these things stemming from an ever-increasing awareness of, and sensitivity to the life-force of planet Earth.

CHAPTER NINETEEN

6.2 Future-research

Summary: A number of concerned people are working together in 'Think Tanks' and International Symposia to seek solutions to the global dilemma.

In September 1967 experts in natural and social science, technology, medicine, and future-research met in Oslo, Norway, for the International Future-Research Inaugural Conference. Future-research was in 1967 no longer new. It had developed, especially in the United States, from its first haphazard beginning in 1944 to a formidable 'think industry', financed mainly for military and related goals, and consisting of 'factories' and 'think tanks' staffed in some cases by well over a thousand 'brains'.

The 1967 Conference was the first of its kind dedicated exclusively to peace and sustained development, and had nothing to do with military needs. The organizers pointed to a new and urgent direction for future-research. Their brief was to establish whether the new intellectual tools of information technology, systems analysis, operational research, forecasting, anticipating, scenario writing and 'future creation' could be used on civilian problems.[1] The most important civilian needs had first to be established in relation to available resources; and the implications of an attempt at 'future-shaping' strategies had to be assessed.

Robert Jungk and Johan Galtung[2] note that very similar questions had been discussed at about the same time among a number of European thinkers. The organizers of the Oslo Conference had hoped it would prove to be a catalyst in furthering this trend. They were not to be disappointed, for out of the conference came a determination that 'the new techniques of inter-disciplinary analysis and prognosis were henceforth to be used against such enemies as urban sprawl, hunger, lack of education and growing alienation.'

The 1967 inaugural meeting could be said to have come of age in the International TIME Environmental Conference of November 1988, more than twenty years later. It had taken a long time for its conclusions to be perceived as urgent international issues, and for the full realization that human and environmental issues could not be separated. But by 1988,

Pace University's Nicholas Robinson could observe that 'On most environmental questions, the nation-state is obsolete ... We have to talk about multinational cooperation'.[3] And John Eddy could insist that 'We've got to get the earth in intensive care, and start to monitor the vital signs of the planet.[4]

By this time Global Warming (the Greenhouse Effect) and destruction of the Ozone layer had both become front-page news. Though scientists still cannot agree on just how much global warming has occurred, to what extent the ozone layer has been destroyed, nor on what the climatic consequences of these developments will be, enough is known to be thoroughly scary. The rates of climatic change have been estimated at 100 times faster than at any previous time in human history.[5]

We know that in ecological matters, the faster things change, the less time there is for evolutionary adaptation, and the more likely it is that there will be serious negative impact. Thomas Livejoy of the Smithsonian Institution points out that there will be no winners in this game of climatic change, 'for it will be fundamentally disruptive and destabilizing, and we can anticipate hordes of environmental refugees dwarfing the numbers of the Dust Bowl era or the boat people'.[6]

Only three years later, the refugee problem from drought, floods and natural disaster had already surpassed even this stern warning. Future-research and action on the lines urged by the TIME Conference had already been more than justified.

By 1988 many people had begun to understand for the first time the stark reality of our predicament, and to accept that nothing less than massive international intervention could stem the tide of environmental destruction. In the absence of global cooperation, man's so-called peaceful 'progress' was turning into a global aggression against the very foundations of life on Earth. This realization was reflected in the very wide range of subjects covered at the Conference, which included, in addition to those on global warming and fluorocarbons, papers on pollution, waste disposal, the population explosion, biodiversity, species extinction, poverty, destruction of the rain forests, environmental protection and economic development.

As in so many spheres where the need for concerted action can be clearly seen, it is economics that has the last word. Thomas Sancton observes that, 'Far more difficult than signing international treaties will be finding the money to make them work'.[7] The impoverished Third World countries in particular, burdened with debt, cannot afford costly environmental projects without outside help. We have to understand clearly and urgently, that financial resources from the industrialized world channelled into sustainable development in these countries, is in no way big-hearted generosity. It is the only means to survival for ourselves. And there are now encouraging signs that this realization is at last beginning to dawn.

A report indicative of this came over the radio-news in July 1991 to the effect that the 'Group of 7', meeting in London to discuss global finance, had included in their agenda the establishment of a fund to save the Amazonian rain-forests. The 'Group of 7' represents the world's seven major industrialized countries. The significance of this move can therefore hardly be over-emphasized, for it shows basic understanding of the issues involved, at the highest possible level of financial debate.

I hope to show that, strange as it may seem, big business is likely to play a major role in the transition to the moneyless economy of the future, as well as in establishing the groundwork for an information-society based on organic principles.

That the new techniques, referred to at the Oslo Conference, should have been brought into being and funded first and foremost for military purposes is a terrible indictment of humanity. At the same time it is an encouraging reminder (of which we are badly in need) that life often uses negative phenomena to positive purposes. One could almost say that this is how life works. The mass death of thousands of beautiful creatures in time of drought works toward the good life for those that survive, and life continually uses excreta and death to fashion new life.

In human terms, tragedy seems to purge the soul and spur mankind to greatness like nothing else. But the price is high. For every emerging hero, thousands are maimed in spirit, and thousands die. We have to move on, I think, and find new paradigms. The Oslo meeting of 1967 was trying to do just this. They did it by establishing relationships. Says Robert Jungk in his preface to the book *Mankind 2000:*[8]

> That almost all people engaged in the struggle for a better and more human future now know each other and have become friends is more than a 'side effect' of the Oslo meeting. In establishing these emotional ties it gave warmth and substance to a community of men and women who think that man is not the victim but the moulder of his destiny. And this confidence in the future and opposition to the darkness of our times may finally prevail.

Today we are still far from achieving the necessary goals, but the work of those engaged in future-research has undoubtedly pointed the way. Hardly a book, or a magazine, or a newspaper, or a newscast but has some reference to that most pertinent question on everyone's minds: If things are really as bad as they seem, what is there that we can do?

CHAPTER TWENTY

6.3 Six Perspectives on Global Problems

Summary: People approach world problems from widely divergent perspectives, but most are agreed that solutions require closely integrated global cooperation.

A new approach emerges in the links between a 'Living Earth', human society, and the micro-chip.

Before going on to explore practical ways and means by which a living information-based society could come into being, I want to take a brief look at some of the suggestions that have been put forward in recent years toward the solution of our most lethal and urgent global crises, looking in particular at how they may agree or differ in principle from those solutions envisaged in my own book in the light of a living Earth. The crucial question is this:

Are these solutions adequate by themselves? If not, are they further illumined by the principles we have found to emerge from a study of living organisms? And if the latter, how are these principles to be incorporated in human society?

We do not lack for suggestions. As indicated in the previous chapter, there's been a lot of hard thinking in recent years and much debate. And out of all this thinking has come an enormous literature pointing to the mortal dangers confronting mankind, as well as on possible directions for change. There has also been the inevitable counter-argument, to the effect that fears for the Earth are over-dramatic, and the technological fix has everything under control.

The literature covers a wide range of disciplines and interests, with emphasis reflecting different perspectives on man's relation to his world. These include:

- *The Ecological Approach*, which sees mankind essentially as part of the environment in which he lives.

- *The Technological Approach,* some of whose adherents accept that by doing 'more and more with less and less', man can transcend the finite limits of Earth's resources, and progress indefinitely towards a technological Utopia. In this view he will also have at his disposal the vast resources of space.

- *The Human Approach*, essentially concerned with the social problems of today's world, by people who remember that they are dealing with human beings, not just with statistics, official bodies and bits of paper.[1]

- *The Political Approach*, which sees problems largely in terms of the balance of power and control of material wealth.

- *The Economic Approach*, in which a sound underpinning of material wealth is seen as the inescapable foundation for social and political stability.

- *The Systemic Approach*, which attempts to combine in one discipline all facets of mankind's life on Earth.

Except for those who have unlimited faith in the technological fix, we shall see that there emerge certain perceptions common to all these thinkers:

- that we and the planet are one and indivisible

- that this unified approach must be expressed in appropriate practical policy

- that we have to find a way to this as a matter of urgency.

Some see it primarily as a matter of spiritual awakening; some as political, economic and social restructuring; some as an ecological crisis. All seem aware that it is in fact all of these things. We have to reach for awakening, apply our new perceptions in the field of human activity, and address the impending ecological tragedy. And we have to do it together. It is *the* basic misapprehension that in matters of energy, food production, markets, inflation, unemployment or wealth, any of us can come out of it without the others.

We now have the capacity, with the help of computers and telecommunications, to draw *everyone* into the system - rich or poor, literate or illiterate - 'good' or 'bad'. The only need is to care. This is the first stage of planetary integration.

What is needed is a synthesis that will enable any person, any family, any region, and any group to become its own source of creation, enmeshed together as a living whole. With the aid of the new communication technologies and the dawning vision of a new world order, I believe it can be done. How far have we got at the present time?

So far as I am able to discover in the available literature, while many have seen far-reaching beneficent implications to the concept of a 'Living

Earth', *no explicit link has yet been made between this concept and the possibilities inherent in the microchip*.

So:

After a brief look at current thinking under the headings enumerated above, we shall be looking at these exciting possibilities in some detail. It is precisely the links between the microchip, human society and a Living Earth which are the subject of my own book.

I shall try to pinpoint those principles of living organisms relevant to human society, and then see to which of them the integrating powers of the microchip could usefully be applied.

CHAPTER TWENTY ONE

6.4 The Ecological Approach

Summary: The so-called 'natural disasters' of the present decade stem in part from man's failure to understand the ecological nature of his world.

The ecological approach has changed radically in the last few years, moving from a situation in which man and nature appeared to be on an inexorable collision course, to one in which mankind and nature were seen as interacting and interdependent parts of the same living system.

This change can only be healthy, because it lays the groundwork for integrated action by all sections of the community in the long-term interest of everyone.

Frederic E. Clements,[1] of The Carnegie Institute, foresaw this trend as long ago as 1939. He noted rather ruefully that while the study of human society was obviously a division of ecology, man's 'importance to himself' would for some time tend to maintain and even to emphasize the existing specialization into sociology, economics, behaviourism, psychology and other fields. The term 'ecology' would therefore tend to be used exclusively for plant and animal life for some years, and would progress toward the direct inclusion of mankind only with the growth of a 'feeling for synthesis'. Students of ecology would continue to be trained in fragmented specialist studies for as long as universities maintained their present structure.

Clements seems to have felt that, while this fragmentation was in one sense greatly to be deplored, it was perhaps a necessary evil, since 'the very essence of ecology is the synthesis derived from the exhaustive analysis of the community and its habitat'.[2] To make this possible, fine-grained specialist studies were essential.

Today, fifty years after Clements' penetrating analysis was written, scientists have accumulated a formidable body of organized specialist knowledge on which the emerging science of community populations can be based. Because of this, and because ecology reflects a synthesis inherent in living systems, we are now moving into an era where the separate sciences converge, giving us a powerful and valid basis for a systemic approach to our problems.

At the end of the nineteenth century, plant-ecology was regarded as physiology carried into the actual habitat. Its theme was one of stimulus and response, a cause-and-effect relation between the habitat on the one hand and the organism and community on the other. Early recognition of man's negative impact on the environment followed this linear cause-and-effect approach, resulting in public perception of ecology as 'nature conservation", and conservationists found themselves trying to save the natural environment from the harmful effects of over-exploitation and pollution, in direct opposition to the forces of economic development. We now recognize the impossibility of separating environment from organism. Life in its entirety constitutes a dynamic integrated system, in which each element acts as an interacting component of the environment of all the others. Man is a valid and vital part of this system.

Clements observes that while cycles in climate, physiography, soil, migration, and abundance bear an organic relation to one another, response takes place at a varying rate and degree, so that 'the mosaic of processes becomes correspondingly intricate,' while 'the best development of ecology demands the synthetic organization of the entire field.'[3]

Clearly this field includes, in the 1990s, the social, political and economic fields of human endeavour, but realization of wider horizons in no way denigrates the invaluable work of the early conservationists, whose prime concern was preservation of the life-support systems of planet Earth. Without them, and the near-panic their work rightly engendered in society at large, we would undoubtedly have proceeded on the headlong road to extinction on which we were set.

Typical of the many dedicated workers in this field were Paul and Ann Ehrlich, whose book, *Extinction*, published by Random House , New York, in 1981, put the conservationist cause in a nutshell.

The gist of this sober and rather frightening book is that during the last fifty years or so humanity has been forcing species and populations to extinction at a rate far beyond nature's capacity to replace them; that in the last twenty five years this process has accelerated alarmingly; and that unless something is done, extinction looms ahead for man too. All this is familiar territory to ecologists, as well as to informed people in related disciplines, but the way in which the Ehrlichs present their material is particularly persuasive and convincing. It is made the more acceptable by the fact that these authors are not just prophets of doom. They do offer alternatives, which is what we are looking for in my own book and in the others under review in this section. But they leave us in no doubt whatever as to the lethal direction of current trends, nor of the urgency of the task to which we should be addressing ourselves.

We are shown how, as nature is progressively impoverished, its ability to provide a moderate climate, recycle wastes, protect crops from pests, replenish soils and so on will be degraded; how an exploding human population leaves less and less space available to the living web of other species, both plant and animal; how these species perform the life-sustaining functions on which, ultimately, we depend; and how, insidious though the destruction of these processes may be, it may cause the collapse of civilization within the lifespan of many people alive today.

The point could be made that there have always been extinctions. The history of evolution was built upon them, and who, after all, would want the dinosaur back? But the Ehrlichs point out that these things happened slowly over a long time span, in an epoch when nature was still capable of replacing lost species with others evolved to do the job. They argue that this is no longer the case, that species are being lost beyond the capacity of the Earth to renew herself, and that, because the biological diversity of species is crucial to the healthy functioning of the planet, we are in very grave trouble indeed. Contemporary extinctions include vast numbers of plants, animals and micro-organisms whose function in the global ecosystem we do not yet begin to understand, and of whose existence we may even be unaware.

We do know that, judging by the rate at which we are destroying their habitat, these species must be vanishing at a rate whose long-term effects are impossible to estimate. In the view of these authors, the rate at which these species are being eradicated has reached the point where a highly industrialized society would be better off doing without even the most 'necessary' projects unless they can be achieved without further extinctions. They point out that species and populations are often endangered far from their habitats, as for instance that a new coal-burning power plant in central Indiana may cause acid rain that will exterminate trout populations in Maine, while interference with larger and more crucial ecosystems can cause catastrophe on a global scale.

The Ehrlichs foretell, as one example of this, the likelihood of rapid changes in global climatic patterns that would follow destruction of the Amazon rain-forests, a perception which has since proved frighteningly true.

> By the end of the century the extinction of perhaps a million species in the Amazon basin could have entrained famines in which a billion human beings perished. And if our species is very unlucky the famines could lead to thermonuclear war, which could extinguish civilization.[4]

That this last observation was no exaggeration was amply borne out by the findings of an historic international conference held in April 1983 and attended by 500 top scientists from all over the world. The conference was sponsored by a formidable list of international scientific, environmental, and population-oriented organizations, and its results made accessible by satellite to United States and Soviet governments. An account of the conference and its conclusions was later published in book form under the title *The Nuclear Winter:The World After Nuclear War.*[5]

In view of the continuing insistence of the major powers on maintaining their nuclear capability, and the determination of many nations to develop it, the conclusions reached at the Conference make the worst kind of science-fiction nightmare look like a tea-party. Clearly we no longer have an option about nuclear war if we want to survive. But the book also reinforces our understanding of the extent to which human life and culture are embedded in, and dependent upon natural processes, and on the need for ecological thinking in all aspects of human life.

It is no longer enough to equate environmental problems with pollution, or conservation, or endangered species. The dependence of human beings upon ecosystem services, which in turn depend on other species, must become part of the thinking of economists, social planners and politicians, just as these aspects of human well-being must be integrated into the ecological paradigm. We have to find ways in which man and ecosystems can interact on a sustainable basis, and the inclusion of human factors in the ecological context does nothing to reduce their urgency. Alarm bells rung by so many early investigators are proving even more urgent than they knew.

Says Thomas A. Sancton[6] in his essay 'Facing up to Impending Disaster', read at the International TIME Conference of 1988:

> Starting at the dawn of the Industrial Revolution, smokestacks have disgorged noxious gases into the atmosphere, factories have dumped toxic wastes into rivers and streams, automobiles have guzzled irreplaceable fossil fuels and fouled the air with their detritus. In the name of progress, forests have been denuded, lakes poisoned with pesticides, underground aquifers pumped dry. ... Perhaps most ominous of all, the destruction of the tropical forests, home to at least half the earth's plant and animal species, continued at a rate equal to one football field a second.

'For decades scientists have warned' said Sancton, 'of the possible consequences of all this profligacy. No one paid much attention.'... 'But', he went on, voicing the feeling of many at the conference, 'this year the earth spoke, like God warning Noah of the deluge. Its message was loud and clear, and suddenlypeople began to listen, to ponder what portents the message held.'

Sancton was not the only one to feel this way. Just about all the papers read at this meeting of international experts reflected their realization of the link between so-called 'natural disaster'and man's careless misuse of his world. Without exception they voiced their sense of urgency, expressed in the outspoken words of Lester Brown, President of the Washington-based Worldwatch Institute: 'We do not have generations, we only have years, in which to attempt to turn things around.'[7]

I have enough faith in mankind to believe that, beginning at last to understand the facts, he will act in time. But it saddens me to think how little of the Wilderness remains, even now.

At about the same time at which I was reading the Ehrlich's book on Extinction, I came upon another, entitled *The Lost Wilderness*,[8] which left me profoundly depressed. The chances for survival seemed so very slight; the process irreversible; the beloved wilderness,

it seemed, already doomed. But in the night, having gone to my bed in a state of near despair, I had a new, more hopeful vision of things. It suddenly seemed possible to link Sheldrake's hypothesis of morphogenetic fields to that of regeneration of the wilderness. In the half-waking exultation in which one tries to hold the conviction of a dream, it seemed to me that since the wilderness had entered so deeply the hearts and minds of men, it would live again despite its present decimation. It even seemed possible that that which had taken millions of years to fashion, would flower again within a little time, *because it had been before and we willed it so*. When we came at last to know ourselves, and to think as one, *we would, of our love, regenerate the wilderness.* And when I was fully awake I could only whisper a prayer that it might be so.

CHAPTER TWENTY TWO

6.5 The Technological Approach

Summary: The most significant technological development of the last few decades is generally acknowledged to be that of the computer. Opinion differs as to whether this is good or bad, but its significance in integrating all aspects of human activity is widely recognized.

The technological approach is one of extremes, and sometimes of unfounded optimism, but where this is the case, it offers a healthy counter to the prevailing gloom, and sometimes we are in need of this. It ranges from a forecast of machines so intelligent (and usually malign) that they surpass human beings and take over the world, to the vision of computer-controlled microscopic devices that operate within living cells to extend human life indefinitely[1] - surely no less horrifying a concept.

Early images of nightmare and breakdown in social organization have to do with the tyranny of the machine, with the suction of men into economic flywheels, and of an inhuman automated existence. However the modern fear, prompted by the vast scale of urban unemployment and the failure, particularly in developing countries, to absorb, feed and house their mushrooming populations, is more closely related to pestilence and starvation, mob violence and social chaos than to mechanised tyranny.

The technology with which I am concerned is specifically that of computers as a means to global communication, and thereby to integration of individual human activity in a happy, functional whole, which sounds innocent enough. But Michael Shallis, who has worked most of his life with computers, agrees with Weizenbaum that computer science is a malign form of modern magic, and presents it seriously as 'the invention of the devil'.[2] This is strong talk. Shallis goes so far as to say that micro-chip technology was born destructive. 'A sane response ... must surely begin with a recognition that the new technology is harmful, is irrelevant, comes from the desire to control and dominate, and is best left alone.[3] Rybczynski, on the other hand, speaks of 'Taming the Tiger[4], and attempts to show that the issue is not so much the control of machines, as control of ourselves, surely a more rational approach. *All* technology is dangerous. Did any sage speak up when our ancestors learned to use fire? But fire was the beginning of civilization. The lens has extended our sight, the radio our hearing, the computer our capacity to think. As such they are extensions of ourselves. They too herald the dawn of a new civilization. But what the new world will be like depends on ourselves. To a greater degree than ever before, it is ours to create.

I have faith in this new world. I believe that sufficient numbers of men and women are caring people, with sufficient benign intent to reflect the essential divinity of life, and that given understanding, we will act aright. My argument rests on this belief. Because of our essential humanity the technological approach is valid, and worth exploring. Its strength lies in its insistence that mankind's future lies in his own hands, and that in matters of human ingenuity the sky is the limit. Its weakness lies in its failure to see man as part of an all-embracing living system, subject to its frailties as well as its strengths, and including in its make-up human aspiration and love.

Perhaps the most glaring weakness of the technological approach is that, of itself, it lacks the substratum of human ethics. Hence the perfectly valid fear that we may unloose on the world 'monsters' we cannot control.

The ecological approach, on the other hand, by virtue of its intricate interweaving of mutually limiting, interactive living beings, has an inescapable, built-in, ethical substructure founded perforce on symbiosis. Arthur Schopenhauer illustrates this to perfection with his tale of the porcupines,[5] who huddle together on a cold winter's day to keep from freezing, but soon feel each other's quills and have to move apart. When the need for warmth brings them close together again, the quills once more force them apart. And so they are driven back and forth 'at the mercy of their discomforts' until they find the distance from each other that provides both a maximum of warmth and a minimum of pain. This kind of mutuality is the very basis of ecology.

In the fiercely competitive industrial world in which we find ourselves, I believe we need both these views. The technologies that gave birth to industrialization undoubtedly let loose a host of harmful environmental effects, but they paved the way for essential later skills.

The two approaches can, and do, converge, and in doing provide a firm foundation for the work of thinkers and futurists. In a partnership between ecological wisdom and the benign use of information technology, I see nothing but good for the human race. Rose-coloured spectacles? Perhaps. But I am not alone in this. There are others who combine a wide view of global affairs with considerable optimism. Conspicuous among them are Alvin Toffler and Jean-Jacques Servan-Schreiber, both of whom see a hopeful future linked to the new technologies and both of whom attempt to integrate them in the wider issues of society.

Alvin Toffler

Toffler is one of the most powerful, far-seeing and articulate of the futurists whose work I have read. Few people can have done more to make us sit up and think, and his controversial books, *Future Shock*,[6] *The Third Wave*,[7] and *Previews and Premises*[8] have deservedly attracted wide attention.

His work covers a very wide field, and so much of it is significant in the organic context that I can only acknowledge its value, and offer here some parts of it which have direct bearing on my own work. It has attracted a good deal of adverse criticism on the basis of its somewhat flamboyant presentation. In spite of this, no-one can afford to disregard these books. They strike at the foundation of many cherished norms and notions, and step on a lot of toes, but their tone is vigorous and cleansing as a mountain stream, their message hopeful, fearless and unequivocal. 'The Third Wave'says Toffler, 'is for those who think the human story, far from ending, has only just begun ...' I'd like to shake him by the hand.

Toffler sees the crumbling of the existing social order as a prelude to a fantastic new way of life, in which our chances of survival will depend on our asking 'the very largest of questions'about our future, and on our capacity to adapt to change. In his book *The Third Wave* he says:

> Humanity faces a quantum leap forward. It faces the deepest social upheaval and creative restructuring of all time. Without clearly recognizing it, we are engaged in building a remarkable new civilization from the ground up ...

This author believes like myself that, even though the decades ahead are likely to be filled with upheavals, turbulence and widespread violence, we will not totally destroy ourselves. He even sees a clearly discernible pattern in the transformations taking place in society, and though he stops short of recognizing them as perhaps the necessary condition for change which my own interpretation implies, he does envisage them as the birth pangs of a new order.

He looks at pollution, corruption, inflation, alienation, loneliness, racism, bureaucratism, divorce, mindless consumerism and other evils as part of the general crisis of industrialism, and observes that economists have warned of a possible collapse of the financial system. In this author's view we are witnessing, in effect, the death throes of industrial civilization. It leaves behind it a world in which, while one quarter of the human species lives in relative affluence, an estimated 800 million people, live in what the World Bank terms 'absolute'poverty.

Since it is precisely this aspect of industrial civilization that became the central preoccupation of my own book, I want to look very briefly at Toffler's vision as to how the new 'Third Wave' forces might be expected to resolve them. He observes that, 'faced by the failure of the Second-Wave strategy, rocked by angry demands by the poor countries for a total overhaul of the global economy, and deeply worried about their own future - the rich nations in the 1970s began to hammer out a new strategy for the poor.'

The new formula called for emphasis on rural development, food-self-sufficiency, and the channeling of resources directly into 'basic human needs.' This formula was what Toffler called 'almost a carbon copy reverse' of the accepted industrial pattern, in which peasants had been drawn into already overburdened cities in search of employment in 'Second Wave' industry. Again in his book *The Third Wave* he says:

> Instead of pushing for labour-saving technologies, the new approach stresses labour-intensive production with low capital, energy and skill requirements. Instead of building giant steel mills or large-scale urban factories, it favours decentralized, small-scale facilities designed for the village.

Advocates of the new strategy had been able to show that many industrial technologies were a disaster when transferred to a poor country, due to the high cost of imported raw materials, the shortage of skilled labour, and a number of interrelated local factors. So what was needed was 'appropriate', 'intermediate' or 'alternative' technology that would fill developmental needs 'somewhere between the sickle and the harvester'.

Many of these ideas stem from the late E.F. Schumacher,[9] and will be more fully discussed later. They had considerable influence on my own work, and are still valid today.

Toffler admits that there is much about this new formula that makes excellent sense: that it confronts the need to slow down the massive migration to the cities; aims to make the villages more livable; is sensitive to ecological factors; stresses the use of cheap local resources rather than expensive imports; and aims to improve the condition of the poor directly, rather than through a 'trickle-down' effect from the capital of the rich.

Yet after all due credit is given - and this is where this author and I part company, he observes that this strategy remains a formula for ameliorating the worst of poverty without ever transforming it. 'It is a Band-Aid, not a cure, and it is perceived in exactly these terms by many governments around the world.'

This viewpoint is only valid if you see the riches of the industrial world as the highest goal of human endeavour. And while it is understandable that poor people should think this way, there are better things ahead if we play our cards well.

It has even been charged that such a strategy might be 'the new form of imperialism', an attitude which Toffler explains by observing that the longer the peasant stays 'down on the farm', so to speak, the less does he intrude on an already overloaded world market, the less does he demand of oil, gas and other scarce resources, and the less troublesome he will be in the political field. In this view intermediate technology and related ideas are subject to the charge that they are 'self-serving' for the rich. Ten years later it is clear that this view dominates the thinking of the Third World, and of other under-privileged peoples.

Meanwhile we are moving towards a set of dynamic new industries based on a convergence of scientific disciplines. These disciplines include in the 1990s quantum electronics, information theory, oceanics, nucleonics, the space sciences and molecular biology, whose convergence, together with the human sciences, reflects the broad interpretation of ecology we've been looking for. It is from this convergence that will spring the networking effect of computers and communication on which my own vision of an organic society rests.

Speaking of the current proliferation of international corporations, and of a range of spontaneous transnational associations that cover virtually the whole field of human interest, Toffler comments on the work of Tony Judge, 'one of our most brilliant organization theorists.' Judge has written extensively about the 'network' character of these emerging organizations of the future, pointing out, among other things, that '*the network is not co-ordinated by anybody; the participating bodies co-ordinate themselves* so that one may speak of auto co-ordination.' This is getting very close indeed to my own notions of a self-regulated, organic society. Toffler adds that:

> What appears to be emerging is neither a corporation-dominated future nor a global government, but a far more complex system similar to the matrix organizations we saw springing up in certain advanced industries ... we are moving towards a world system composed of units densely interrelated like the neurons in a brain rather than organized like the departments of a bureaucracy ...

Just how close, I find myself wondering, can authors like these move towards the concept of a living Earth without actually getting there?

Toffler notes that the more diverse a civilization becomes (the more differentiated its technology, energy, forms, and people), the more information must flow between its constituent parts if the entirety is to hold together, and this too is precisely what we find in living systems.

As a result, people and organizations require more and more information, and 'the entire system begins to pulse with higher and higher flows of data.'

All this ties in very well with my own notions of self-governance, and in particular with the ways in which the new technologies can help in resolving the plight of the very poor - my perennial obsession - but perhaps Toffler comes closest to my own approach in his refusal to see the contemporary scene in anything but hopeful terms, and his certainty that mankind has the power and the will to build himself a new and better world.

Jean-Jacques Servan-Schreiber

Most dynamic of all in this line of thinking is perhaps the French writer Jean Jacques Servan-Schreiber, whose book *The World Challenge*[10] came out in English translation in 1981.

This book was of particular interest to me, because its author shares my belief that the key to global integration lies in computers and communication. It is by far the most exciting book I've come across on the application of micro-electronics to global affairs. Based on the work of the so-called 'Paris Group'which was formed in the summer of 1979 to analyse world problems, and to which Servan-Schreiber himself belonged, it describes the attempt of the group, assisted by top economists, politicians and scientists from Japan, the United States, the Arabian Gulf and the Third World, to find an intelligent way towards world sanity.

What emerged from the meetings of the Paris Group was the conviction that mankind is finally within reach of the only infinite resource - information, the growth of human minds - and that the means to world-wide integration, now well within our grasp, was electronic communication.

Servan-Schreiber notes that the work of the Group was given substance and credibility by the strong participation of Japan, whose meteoric rise to industrial supremacy since World War II is legendary, and whose grasp of the steps necessary to stabilize the world economy is way ahead of most of us.

The essence of the Japanese plan to check the disintegration of the world's economy was a partnership between all advanced countries, to extend the benefits of the new technical era to everyone. The industrial giants could no longer concentrate on expanding overseas markets, but must become closely integrated as one world.

Basic to this perception was the knowledge that oil, the basic source of energy, was no longer cheap and inexhaustible, and that the entire industrial base of the world must be rebuilt with another kind of raw material: the microchip. The chip would eventually replace the barrel of oil as the basis for a new kind of society based on the communication of information, 'with one crucial range of development, one basic raw material: the trained human brain.'

Servan-Schreiber points out that this time the Japanese knew they must share their knowledge with others; that the system established at the end of the war had put the Third World at such a serious disadvantage that it demanded correction in depth ... and that the only way to come to terms with it was a massive transfer of technology to the Third World. This would make the training and development of human resources the priority of all priorities.

The Paris Group concurred that the only transfer that counts, the one that all development will depend upon, is each person's move from the stage where the means of learning are lacking, to the stage where he or she can develop his or her own faculties, abilities and creativity.

What is needed to begin such a transfer to huge impoverished countries had no resemblance, in this view, to the huge steel and concrete equipment characteristic of classical industrialization. What was needed was simply the extension of the soft infra-structure of computerized society; 'the means that link every workplace, in every location, to a worldwide network of knowledge, the means ... that disseminates the elements necessary for on-the-spot training of people according to their needs.'[11]

This approach throws an interesting side-light on the vexed question of unemployment, to which we will return later. The Japanese Mitsubishi report to the Paris Group insists that 'in the considerable field of ... industrial activities that are going to be computerized, in the field of 'software', microprocessors and robots, if we employ the five billion inhabitants of the planet, we will still be short of personnel.'

Servan-Schreiber observes that the only brake on full implementation of computers in a functional information society is our own incapacity. Quoting again from the Mitsubishi report, he notes that 'the capacity to produce instruction and programs, which is the realm of man himself, has fallen far behind the information system's physical capacity to obey.' In this view, confirmed by experience, we are already finding that the larger the number of microprocessors, the greater the need of a vast human contribution.

> Each computerized function requires ten, one hundred, a thousand brains to compose, formulate, imagine and design the task of the programming environment, which can take root only in an increasingly developed and creative society.[12]

Ten years after this book was written we are discovering to our cost that it is not only programming, but skilled maintenance of our machines that is hampering their wider usage. So once again it is a question of people, the full employment of people, not only in terms of a job for everyone, but also, in Servan-Schreiber's memorable words, 'in the employment of each person's full faculties.'

My own view of this encouraging vision is that human fulfillment requires the integration of many diverse faculties and skills beside those of our minds, and that the prime role of computers and communication will be to make such integration possible. Just how this can be achieved will be discussed at length in later sections of this book, but my reservation only adds another dimension to a brave and far-reaching endeavour. A society in which the new technologies will help to offer *not only full employment, but the full employment of each individual's human capacities* is as good a dream as any we might have.

CHAPTER TWENTY THREE

6.6 The Human Approach

Summary: Work by E. F. Schumacher, by Peter Wilsher with Rosemary Righter, and by James Hemming illustrate the Human Approach to social problems in the fields of labour, self-help and human motivation respectively. Emphasis is in each case on the value of the individual man.

Three Aspects Of The Human Approach

The human approach to global affairs covers so wide a range of perspectives, both individual and social, that I can only hope to touch very briefly on those most closely related to my own work. These are:

- The individual in his place of work, as exemplified in the writings of the late E.F. Schumacher.[1]

- The plight of the very poor, given graphic and sympathetic treatment in the work of Peter Wilsher and Rosemary Righter.[2]

- Individual morality and human motivation, in which the writings of James Hemming[3] closely parallel my own beliefs.

Looking closely at other people's work is the only way I know of assessing how much of what I perceive as truth is really mine; of grasping the most significant trends in contemporary thought; and of trying to lay the foundations for something new. So while looking at the aspects of current thinking listed above, I will try to put them in perspective in the context of my own work.

I have chosen these particular books because their essence in every case is a feeling for people. Their emphasis is on the vastly underrated capacities of the ordinary man; their faith that given hope and courage, and the opportunity to help himself, man is invincible.

E.F. Schumacher

To start with, then, man in relation to his work. This aspect of the human approach to social organization is given sane and essentially humane treatment in E.F. Schumacher's now famous book, *Small is Beautiful.*

Schumacher was concerned with people as individuals. Not with aggregates of people as 'labour', but with the meaning, to each individual man, of earning his bread in a heavily industrialized society.

This approach is rather a joy to a layman like me, because, alone among economists, Schumacher avoids specialized economic jargon and confines himself to language any of us can understand. He does not talk about 'externalities', or 'marginal production', or 'the postulate of non-satiety'. He talks about economics 'as though people mattered'.[1] His essential concern is with the poor; with the day to day experience of ordinary men without motivation or hope. 'It is a strange phenomenon indeed', says Schumacher, 'that conventional wisdom of present-day economics can do nothing to help the poor.'

In point of fact it isn't really strange. Current economic thinking still has its feet in an earlier age. In keeping with other disciplines, the time is ripe in economic thinking for a whole new paradigm, a process which Schumacher himself has done much to bring about.

Schumacher suggests that what is presently taught as economics by-passes the poor, the very people for whom development is really needed; that the economics of giantism and automation is a left-over of nineteenth century thinking and is totally incapable of solving any of the real problems of today; that an entirely new system of thought is needed which shall be based on attention to people rather than to goods; and that it has become unbelievably urgent in the twentieth century consciously to use our enormous technological and scientific potential in the fight against misery and human degradation.

This can only be done, in Schumacher's view, in intimate contact with people - with individuals, with families, with small groups, which presupposes a political and organizational structure that can provide this intimacy.

This is so close to my own view that I could wish most sincerely to have written it myself, and though this author lived and worked just ahead of the startling developments in computer technology on which my own solutions depend, his vision is no less valid today.

In taking notes from Schumacher I kept finding bits that I wanted to underline because they overlapped my own thinking and seemed so important. But of course they seemed important precisely *because* they matched my own thinking so closely.

Because people can be themselves only as individuals or in small groups to whom they can relate, we must learn to think in terms of an articulated structure that can cope with a multiplicity of small-scale units. Schumacher insists that unless economic thinking can grasp this, it is absolutely useless. Unless it can make real contact with the human realities of poverty and despair, it should be abandoned and we should start afresh.

This, I realize, is exactly what I am trying to do. It wasn't yet quite possible when Schumacher wrote. With the help of computers it is so today. I like this author's courageous, open-minded approach. It is exactly what is needed at a time when a fresh start has never been more urgent nor its direction more obscure.

Because of his deep concern for the poor, Schumacher gives considerable attention to the relationship between wealth and human labour. He suggests that in the view of the modern economist, 'labour' is little more than a necessary evil. From the viewpoint of the employer, it is simply an item of cost, to be reduced to a minimum if it cannot be eliminated altogether, say by automation. To the workman, on the other hand, to work is to make a sacrifice of one's leisure and comfort, and wages are a kind of compensation for the sacrifice. Schumacher points out in his usual fresh and delightful way that hence 'the ideal from the point of view of the employer is to have output without employees, and the ideal from the point of view of the employee is to have income without employment,' a remark that seems to me to pop many of our Western employment problems neatly into the proverbial nutshell!

As opposed to this situation, Schumacher cites the point of view of the Buddhist economist, who regards labour in an entirely different light. In the Buddhist view, work has a threefold function: to develop a man's individual faculties; to give him the chance to cooperate with other people in the performance of a common task; and finally, to produce just so much in the way of goods and services as will provide a 'becoming' existence for each man. This is indeed a very far cry from our Western concept as to the purpose of labour, in which 'output' has held predominant place throughout the history of economic theory. Schumacher observes that:

> ... to organize work in such a manner that it becomes meaningless, boring, stultifying, or nerve-racking for the worker would be little short of criminal; it would indicate a greater concern with goods than with people, an evil lack of compassion, and a soul-destroying degree of attachment to the most primitive side of this worldly existence.

I confess to some puzzlement as to why, if there are communities actually living by these lofty ideals, we have not heard of them, but

Westerners like myself are, with some notable exceptions, proverbially ignorant as to how Easterners live. In themselves such ideas are faultless.

Schumacher points out that if a man has no opportunity to obtain work, he lacks both an income and the motivating force which disciplined work can provide, and is in desperation. When a modern economist engages in sophisticated calculations as to whether full employment 'pays', such desperation does not figure among his variables. The point at issue is whether it might not be more 'economic' to run an economy at less than full employment 'in order to ensure a greater mobility of labour,'a better stability of wages, and so forth. Translated, this means: 'in order to hold wages down by keeping people on the move till they get desperate' - but Schumacher does not put it this way. He notes that in economic thinking the fundamental criterion of success is simply the total quantity of goods produced during a given period of time.

> From a Buddhist point of view, this is standing truth on its head by considering goods as more important than people and consumption as more important than creative activity.

Expanding on this idea Schumacher explains that since consumption is merely a means to human well-being, the aim should be to obtain the maximum of well-being with the minimum of consumption - a notion clearly in line with the need to cherish and conserve global resources, and with the management of the land itself.

Schumacher notes that the land is regarded, in the Buddhist view, as a priceless asset which it is man's task and happiness to 'dress and to keep'. This presupposes three goals: health, beauty and permanence. Thus agriculture must fulfill at least three tasks:

> - keep man in touch with living nature, of which he is and remains a highly vulnerable part.
>
> - humanize and ennoble man's wider habitat.
>
> - bring forth the foodstuffs and other materials which are needed for a becoming life.

Schumacher voices the conviction that a civilization which recognizes the third of these tasks, and which pursues it with such ruthlessness and violence that the other two are not merely neglected but systematically counteracted, has no chance of long-term survival. This statement projects his criticism of Western economic thinking into the very centre of the current ecological debate.

The title of Schumacher's book, *Small is Beautiful*, stems from his conviction that technology should be geared to lead us back to the real needs of man, 'and that also means to *the actual size of man*. Man is small and therefore, small is beautiful. To go for giantism is to go for self-destruction.'

In applying these ideas to the needs of developing nations, Schumacher expresses the belief that the first need is to start work of some kind that brings individual reward, however small, since it is only when a man experiences the immediate value of his labour that he becomes interested in making it more valuable still.

> It is therefore more important that everybody should produce something than that a few people should produce a great deal ...

In this view the task confronting developers is to bring into existence millions of new workplaces in the rural areas and small towns, a task that heavy industry cannot perform.

Schumacher suggests that the real task before us may be formulated as follows:

> First, that workplaces have to be created in the areas where the people are living now, and not primarily in metropolitan areas into which they tend to migrate.
>
> Second, that these workplaces must be, on average, cheap enough so that they can be created in large numbers without this calling for an unattainable level of capital formation and imports.
>
> Third, that the production methods employed must be relatively simple, so that the demands for high skills are minimized, not only in the production process itself, but also in matters of organization, raw material supply, training, marketing and so forth.
>
> Fourth, that production should be mainly from local materials and mainly for local use.

In Schumacher's view these four requirements can be met only if there is a 'regional' approach to development - a theme given considerable attention later in my own book - and also if there is a conscious effort to develop and apply what he calls an 'intermediate technology'. This last entails methods and equipment which are:

- cheap enough so that they are accessible to virtually everyone
- suitable for small-scale application

- compatible with man's need for creativity.

We have seen that, in spite of the obvious relevance of Schumacher's views to the well-being of poverty-stricken peoples, and in spite of the explosive proliferation of intermediate technology that followed publication of his book, intermediate technology has, in some quarters, become politically suspect. This is because it is possible to read it as an attempt to keep developing nations from catching up with those already highly industrialized. I think this is sad. If ever there was real concern for the poor, it was surely to be found in this man, and while in today's world we shall always need certain large-scale corporate concerns only made possible by the highly sophisticated techniques of modern heavy industry, the plight of the very poor cries out for action at an immediate and more personal level.

It is at this level that Schumacher's work relates most closely to my own search, both in concern for the poor and concern for the planet, and though the new electronic technologies had not developed in Schumacher's day to the point where they are now, he was vividly aware that the best kind of aid to give to any human being was a gift of knowledge that would enable him to help himself. Free access to all available knowledge by means of computers and telecommunications is thus entirely compatible with his ideas.

Schumacher insists throughout his work that the ever-increasing preoccupation of aid programmes should be to make men self-reliant and independent by the generous supply of the appropriate intellectual gifts, gifts of relevant knowledge on the methods of self-help.

My own view precludes the need for 'aid' programmes in the sense in which we have been used to thinking of them. I would rather say that the ever-increasing preoccupation of all social organization should be to make men self-reliant and independent, not by the 'generous supply' of intellectual gifts, but by their free and unrestricted interchange. Who, after all, is equipped to know what might be appropriate for whom? Schumacher would surely agree, were he here to tell us so, that using the new technologies to free all knowledge to all men is the greatest gift of all.

Peter Wilsher and Rosemary Righter

The best human approach to the social problems of poverty, unemployment and homelessness I have come across is that of *The Exploding Cities*,[4] by Peter Wilsher and Rosemary Righter. This sensitive and sympathetic study deals in particular with urban squatting in the Third World - the cities the poor build - and can best be contained in their conviction that the key to improvement lies in the initiative and tenacity of the squatters themselves:

> The only long-term policy must be to help the teeming, toiling millions of the Third World cities to do as much as they can, and want, to help themselves. Otherwise the pessimists in the end will turn out to be right - 14,677 million desperate and dispossessed people will use their physical and political power to help themselves in other, far more cruel and dangerous ways.

Writing in 1975, they suggest that of the 2,000 - 3,000 million expected increase in human population during the next three decades, three out of every five can be expected to join the teeming hordes of those Asian, African and Latin American cities which will have most difficulty in feeding them, housing them, transporting them and finding them jobs - and which are already, in many cases, bursting at the seams (not to mention the sewers, the water supply, the tramway system and most other aspects of civic life).

> Calcutta is of course the awful example: the suppurating image of what all these other cities - maybe the whole world - could so easily become, under the inexorable influx of too much humankind.

Conventional accounts of urban squatting, say these authors, suggest that urban problems are the consequence of an unassimilated influx of incapable people at the margins of society who are a drag on modernisation. From long experience of living and working among them, Wilsher and Righter assert that nothing could be further from the truth. Their experience has shown that it is very often those with a high degree of intelligence and initiative who have braved the daunting challenges of urban poverty.

In this view, since illiterate and disadvantaged people often cope with appalling living conditions with a surprising degree of competence and adaptability while maintaining some semblance of family life, effective change depends on permitting and encouraging a high degree of self-help wherever possible.

Low-income owner-builders all use resources that no large or official organiaztion can employ. In all places where they are free to build, they employ diverse resources, including small quantities of exceptionally cheap materials, equipment and help that can be borrowed from friends and relatives. Only when building is on a small scale can full use be made of local conditions and resources, and these individually small but collectively vast resources far outweigh those that are possessed and directly controlled by large commercial or state enterprises.

These small-scale buildings are developed in response to family circumstances rather than to professional or governmental planning, and result in a social flexibility which is of great importance to poor families. Wilsher and Righter say that there are many recent studies to confirm the benefits of the fine-grain cities created by local entrepreneurship in housing, and that they give low-income people in particular a far greater opportunity to match the places they live , the dwellings they live in and the tenure they have secured with their sensitive and changing priorities.

Central authorities can support local housing action groups in two essential ways: by making basic resources available, or by providing infrastructure services that individuals cannot provide for themselves. But when governments decide for people and impose their own institutional solutions, and especially where these involve mass removals to greatly improved surroundings, their well-meaning attempts often result in a lesser quality of life for the poor, through loss of immediate small-scale job opportunities close to where they live, through transport costs to and from places of work, and from the loss of access to the intricate web of lateral information and exchange on which informal communities operate.

It seems clear that whether we are looking at informal economic activities or local housing, it is their flexibility and adaptability to local and personal needs and opportunities that make them so effective and necessary for economic and social development. In this way Wilsher and Righter's emphasis on the 'vastly underrated' capacities of the ordinary man is entirely in line with my own search: a synthesis that will enable any person, any family, any region, and any group to become its own source of creation. But there are serious limitations to these informal activities while we remain within the present organizational system, and while our numbers continue to escalate at such an alarming rate - limitations of which these authors are very well aware:

> 'But having said all that, and giving human beings the maximum credit for toughness, tenacity, survival-power, ambition, and bootstrap-raising ability, the fundamental question remains: can the people and institutions ostensibly in charge actually cope with the mountingly chaotic urban ant-heaps?'

The question could equally well be asked of the whole economic, social and political scene at the present time. And the answer is clearly 'no'. We *have* to find our way to newer dreams.

James Hemming

Everyone wants a world in which human well-being is maximized. 'But,' says James Hemming, in his fine book *Individual Morality*,[5] 'that is the start; not the finish.' The start of what? No-one knows.

But, says Hemming again, 'our act of faith must be that we believe in what happens when human vitality and creativity are released'. I endorse that faith. But how to release the best and creative in man?

We like to think of humanity as God's highest creation. We like men and women to go on their way happy and creative, reverent and fulfilled. Instead we live in a world of political rivalries and pretensions, of hypocrisy and cruelty and despair. We live in a world most of whose citizens exist in abject poverty. And though the Mother Theresas of this world go as a shining light amongst us, their heroism does little to change our world.

People do care. Benevolent 'causes' numerous as the stars demand our support. But they are diverse, and sometimes conflicting in their approach. Each seems deserving of personal support, yet none is sufficient, of itself, to cover the urgency of all. We need an overall commitment to give meaning to our lives, for in the chaos of our daily more tumultuous world, nothing short of total commitment can meet our personal need.

Hemming observes that some people are content to find meaning at the level of immediate personal relationships. If their family and social relationships are happy, their work satisfying and their financial position adequate, they will be content for as long as these things last. But he notes that some people need to extend their relatedness and sense of value beyond the most intimate circle of family, of friends, or of personal objectives:

> They need the assurance that their own lives contribute to fulfillment of the wider concerns of mankind.

One has to add that in today's world personal satisfactions seldom last very long. And when they fail, unless we are buoyed by a deep religious faith not given to many, the bottom falls out of our world. We need much deeper roots than these. We need a meaning transcending our personal lives or the personal salvation of our souls.

Twenty years ago, when Hemming was writing his book, traditional values everywhere were breaking down, and ecological concerns were in their infancy. He observed that for those who sought a wider relationship, it was no longer clear what, if anything, warranted personal commitment. Yet he clearly foresaw what such a commitment must be.

> Our primary commitment is to life and to the Earth and to one another. No other value can now satisfy us.

To me this seems exactly right. Given such a commitment, any sphere of activity in which we find ourselves will converge in the good of the whole. To salvage and maintain our world will employ at fullest stretch our deepest understanding; our most profound morality.

Hemming talks of:

> ... certain principles of being and relationship that are as pragmatic and inevitable as the laws of physics ... Before we can begin to frame a dynamic moral order out of pragmatic principles, we have to clear away a very ancient confusion - that the wishes of the self and the needs of society are inexorably in conflict.

Here again Hemming foreshadows my own perception that a healthy self-interest within the framework of wider commitment is a sound biological principle. He suggests that we are today urgently in need of a common ground of awareness, 'which can give unity and direction to our open society, without destroying the rich and valuable variety it engenders'. This perception too is in line with the organic view of mankind, and will emerge time and time again in my own search. If Earth is the living creature I think she is, we do not have to look further for common ground.

CHAPTER TWENTY FOUR

6.7 The Political Approach

Summary: The political organization of human affairs belongs to the past, and is no longer tenable. The structure of society must now evolve towards the systemic principles found in living organisms.

I am going to stick my neck out and say that so far as I have been able to discover, nobody has the faintest idea how to solve global problems politically, for the simple reason that it can no longer be done. 'The dynamic moment is elsewhere, in the difficult business of gaining confidence in our own energies and capacities.'[1] We have to find better ways in which the individual contribution can merge with the collective experience, remembering always that it is the quality of the individual that generates the nature of the nascent collective. If each individual did not matter intensely, we would sink into mediocrity.

Listening to the bitterness and wrangles of today's political world, one is forced to the conclusion that the procedures are far too crude to cope with the intricate mesh of social and economic complexities in which we are embroiled, and in which personal and national self-interest of necessity take precedence over global commitment. It is impossible to extricate politics from economic considerations for so long as economic viability depends on competition between individuals, companies and nations. By the same token it is impossible to keep policies free from bias and corruption while monetary advantage persists.

Having said that, and with the sword of Damocles now hanging over my head, what do politics really mean? If politics cannot serve to resolve our problems, are they still useful? In as much as political science is the science and art of internal government and international relations, it is clearly paramount in human affairs so long as the nation-state survives. But can the nation-state survive?

We are in the midst of the most profound political and economic upheaval the world has ever known, and no-one can say where it will lead. But one stark fact stands out: Most of the world lives in abject poverty. Unless politico/economic systems can help the poor, they are doomed. We haven't found one yet that can meet the bill.

We have today about two-thirds of mankind living - if it can be called living - on less than 30 cents a day. We have today a situation where there are about one billion illiterate people around the world, although the world has both the means and technology to spread education. We have nearly 70% of the children of the Third World suffering from malnutrition although the world has the resources to feed them. We have maldistribution of the world's resources on a scale where the industrialized countries are consuming about twenty times more of the resources per capita than the poor countries. We have a situation where, in the Third World, millions of people toil under a broiling sun from morning till dusk for miserable rewards and premature death without ever discovering the reasons why.[2]

This intolerable situation and the angry response of Third World leaders in international fora have given rise to brave attempts by the United Nations and world leaders to formulate international solutions to the problems of poverty. The Third World has long demanded radical change in a political and economic system which, it contends, systematically discriminates against its interests. It insists on fundamental structural change that will redress past patterns of hopeless dependency and provide real opportunities for a share in global growth, and this insistence grows increasingly vocal in United Nations debate.

Some of these demands found expression in the resolutions of the Sixth and Seventh Special Sessions of the United Nations[3] nearly twenty years ago. They include:

I. International Trade.

- Complete reorganization of the terms of international

trade, with adequate international financing facilities, and favourable market arrangements for Third World countries that will help to redress past imbalances.

II. Transfer of Real Resources

- A transfer of real resources from developed to developing countries, including substantial and sustained increase in financial aid amounting to 0.7%of their gross national product no later than the year 1980.

- A link between special drawing rights on the International Monetary Fund and development assistance to Third World countries.

- A substantial increase in the resources of the International Development Association to be used for the assistance of underdeveloped countries.

- The largest possible share in the procurement of equipment and consultancy services.

- Mitigation for the debt burden of developing countries.

- Better access to the capital market of developed countries.

III. Science and Technology

- Cooperation by the industrial world in the establishment, strengthening and development of the scientific and technological infrastructure of developing countries.

- Direct assistance to developing countries in their science and technology programmes, and an increase in the research and development programmes of the developed world which might be of interest to developing countries.

- An international code of conduct for the transfer of technology, corresponding to the special needs of the developing countries.

- The adoption of policies to counter the 'brain drain'from developing countries.

IV. Industrialization

- Encouragement of investment projects in developing countries consistent with their needs.

V. Food and Agriculture

- Assistance to developing countries for agriculture and food production.

- Increase in World Food Security through the build-up of world food-grain reserves, large enough to cover foreseeable major production shortfalls.

The extent to which politics and economics are interlinked could hardly be more clearly enunciated than in these demands. The basic premise underlying them is the desire of Third World countries to follow the same patterns of development as those of the industrial West, a pattern only possible when the world population was a great deal smaller than it is today. Global resources simply cannot sustain this kind of growth.

There has been much talk, in certain quarters, to the effect that there are in fact no such material limits. Gross increase in world food production is cited in support of this view, while the environmental cost of present 'advanced' methods of agriculture are brushed aside. But even were one to accept such arguments at face value, there are social limits to growth as implacable as the material ones, a subject most ably dealt with in Fred Hirsch's book *Social Limits to Growth* published by Routledge and Kegan Paul in 1977.

The heart of Hirsch's argument is that once the mass of the population has satisfied its main biological needs for life-sustaining food, shelter and clothing, attention moves to less tangible social demands which are in essence self-limiting. One or two simple examples will illustrate the point. A man's desire for a high-powered car and good open roads on which to drive it can only be satisfied while relatively few people use the roads. Thereafter he will drive slowly, bumper to bumper. Or a man's love for the quiet and privacy of natural surroundings will avail him little once too many people have a liking for the same beautiful spot and converge upon it. Too many people set limits to our options.

Our central problem, then, is demographic in essence rather than political, but while urgent action is clearly called for, it is hardly surprising that a demand based on the twin concepts of economic transfer and indemnification did not evoke enthusiastic response in world fora.

Massive and increasing 'aid' of one kind and another has continued through the years, but often with disappointing results. Difficulties of distribution are one contributing factor. Another is the lack of infra-structure and technical expertise in developing countries. Another the sheer magnitude of the task. Political elites in Third World countries, usually trained to a large extent in the West, discover that Western standards are impossible to apply in their countries on any mass scale, and tend to curl in, concentrating instead on personal wealth and power.

While one cannot but have sympathy for the dilemma of leaders struggling with an impossible task, respect for the individual insists that no lasting good can be expected to come to a nation, save from the initiative, hard work and vision of its people themselves. Tinbergen suggests that the Third world is not demanding massive redistribution of wealth. It is rightly asking for equality of opportunity and insisting on the right to share in future growth. I'm afraid this modest and essentially just demand is not borne out by the recommendations of the U.N. Sixth and Seventh Special Sessions mentioned above, and while untold millions have been poured into developing countries since Tinbergen wrote, a total restructuring of the international order has seemed until recently rather far off the mark.

The collapse of communism in Eastern Europe and the Soviet Union, and the current demand for independence in an ever-growing number of states, presents a whole new ball-game in the politics of the 1990s, but just how far this new orientation can go in solving the problems of widespread starvation has yet to emerge.

My own view is that we have to find a wholly different road to prosperity, based, as I hope to explain more fully, on organic principles - a prosperity based on the employment of each man's fullest capacity, and made possible through the role of computerized organization. When information and the human mind become the prime resource, 'wealth' as we know it today will rapidly become as obsolete as the nation state.

While the supply of food to starving people remains a clear moral imperative, the only really useful long-term transfer to the Third World will then be the new technology, and access to information, with emphasis on those procedures consistent with both human and planetary well-being. Thereafter it will be a wholly different scene. Basic necessities will flow to all as in any other living body, and opportunity will be open to all, but the quality of individual life will depend on oneself alone.

The means to transition towards this kind of global integration will be discussed at some length later in this book, but already it seems clear that progress can only stem from the combination of knowledge, the enquiring mind, and a sturdy, energetic self-reliance. These, as it seems to me, are the least that society will demand from us.

If one looks up the word politics in a good dictionary, one finds that it's secondary meaning is:

> ... that branch of moral philosophy dealing with the state or social organism as a whole.[4]

In this last broader sense of the social organism, it is perfectly applicable in any truly global context.

Lewis Thomas speaks of the modern nation as 'probably the most stupefying example of biological error since the age of the great reptiles.[5] I believe that, as a political, economic and military unit, the nation-state will die like the dinosaur in the foreseeable future, if for no other reason than that the role it was designed to play is no longer relevant. However, in the culture of peoples the situation is quite otherwise, and the validity of genetic and cultural affinities is in no way under dispute in this discussion. I would go so far as to say that the differences they engender are of the essence, and will remain so.

If national governments are giving way in the regulation of human society to other more flexible norms, who will make the laws by which we live? Are there values on which laws could be based applicable to humanity anywhere? I believe that there are.

We know that in different cultures there are widely differing concepts of excellence and a wide variety of behavioural traditions resulting therefrom. But certain qualities such as courage and kindness are revered throughout human society. It is universal values like these that must be the basis of our search.

But are they specific enough to form a basis for universal law? A paraphrase of the words of Senior Council Ishmail Mahommed, Judge of Appeal in Lesotho and Swaziland [6], goes something like this:

Laws should be *relevant*, and perceived to be relevant, to the ordinary man, i.e. they must be part of his contemporary emotional experience.

Laws should be *legitimate,* i.e. laws must be perceived to be more than a set of rules designed to serve the ends of particular groups or a particular race.

Laws should be perceived as having sovereignty and a purpose that go beyond the power of those that execute them. Rulers come and go. Governments come and go. But the law has an enduring capacity, an enduring mission, to perpetuate and to further justice. And justice is native to the soul and the spirit of man himself - beyond the idiosyncrasies of a particular culture, beyond the idiosyncrasies of a particular historical period, beyond the idiosyncrasies of a particular ruler - because ultimately, without the sovereignty of the law, there can be no civilization whatever.

I find myself wondering whether justice is really this kind of ultimate. I'm not sure. It is tied to the notion of 'rights' which I find so disturbing. Justice implies that things should be fair. I don't think they can be. What is fair about a child being born blind because it parents have syphilis, or a child being born with AIDS? Or that a difficult birth should result in mental retardation of one child while another has all his faculties intact? Even when equal opportunity is afforded to both, the one with superior mental powers will inevitably end up in a better situation than his sibling. Surely we now need to move toward ideas about devotion and responsibility rather than rights?

Sovereignty of the law is of course something else again. I can't seem to link it to justice so much as to the common good. After all an eye for an eye could be conceived as just, could it not? But it is not ethically acceptable. Nor can it be allowed if society is to find its way to peace of mind. Only in the sense of equal opportunity before the law is justice tenable, and this is what this speaker presumably means.

Even the concept of equal opportunity seems to present pitfalls when we try to embody it in national law. In an article in *The Citizen*[7] of 13 December 1989, Lars Foyen reports that Sweden's dream of creating equal opportunity for all through an ambitious state school system is turning into a costly nightmare. After decades of investment, it is not working.

The Social Democratic government says it spends more on primary and secondary education than any other country and its commitment has been the envy of educationalists around the world. But teachers have gone on strike. Pupils have demonstrated in protest against dilapidated buildings and outdated books. And an official survey shows that equal educational opportunity has done nothing to change hidebound class patterns. Absenteeism is on the rise, and educational standards are deteriorating. Thousands of children leave school without being able to read and write properly. As in so many situations in which a wide variety of factors contribute to the end result, results have been different from expectation.

When you apply notions of fairness to the complexities of international concerns, and attempt to legislate them, the desired results are even more difficult to attain. I suggest in all humility that self-organization such as we find in a living body is the only way out of this dilemma.

CHAPTER TWENTY FIVE

6.8 The Economic Approach

Summary: New concepts are introduced into economic thinking by Herman Daly and Hazel Henderson.

The two writers on economic affairs whose work bears most closely on my own are Herman Daly and Hazel Henderson. Their books, *Toward a Steady State Economy*[1] and *The Politics of the Solar Age*[2] respectively, both break away from conventional economic theory, and offer valuable pointers toward a sustainable Earth.

Daly asks three simple questions:

- How to live on a finite earth?

- How to live a good life on a finite earth?

- How to live a good life on a finite earth *at peace* and without destructive mismatches.

Put like that the issues sound so simple that I found myself wondering, when reading the book, whether perhaps the answers would prove equally simple, if we could only see them. I believe now that they are. By the application of one unifying technique - the free flow of information - so many of our problems simply melt away.

Daly ponders as to why these all-important questions do not constitute the core of economic theory, and concludes that before this can happen economists must undergo a revolutionary paradigm shift and sacrifice large intellectual vested interests in the perpetual growth theories and policies of the last thirty years before they can really come to grips with these questions. He notes that the physical scientists, unlike economists, 'are viscerally convinced that the world is a finite, open system at balance in a steady state, and they have not all invested time and energy in economic growth models.'

I am aware of the total inadequacy, and perhaps of impertinence in extracting like this from the work of thinkers, in this case of economists, about whose discipline I know almost nothing. I have tried to avoid the worst pitfalls by quoting their work wherever possible rather than discussing it, hoping that, in spite of its fragmentation, the pieces will hang together sufficiently well to be self-explanatory in the context of this book without doing their authors an injustice. The importance of what these writers are saying is my only justification, and I can do no better than to pay them tribute while craving their indulgence.

As I understand it, then, Daly suggests that economics should be considered as a steady-state open system similar to that of the earth itself, and of other organisms. He suggests that economists, in pursuing the concept of 'value', early lost sight of the fact that what he calls 'its physical dimensions' (i.e. the actual products of the earth on which the economy rests) do not have the capacity of infinite growth. We are therefore constrained to keep 'growth' within the limits of a steady state, in line with the capacity of the earth to maintain itself. Whereas the concept of economic growth has perhaps no perceivable limits in terms of 'value', the same is not true of the physical products - the 'mass' on which that growth finally depends.

Sustained increase in the Gross National Product (G.N.P.) of a country has long been the prime goal of economists throughout the world, and is still seen by most governments as the prime indication of a healthy economy. Daly's work breaks new ground in pointing out that there are serious flaws in this view, since the needs of a developed country are quite different from those of a country in the early stages of economic growth.

> [Growth] in G.N.P. in poor countries means more food, clothing, shelter, basic education and security, whereas for the rich countries it means more electric toothbrushes, yet another brand of cigarettes, more tension and insecurity, and more force-feeding through advertising.

The upshot of these differences is that, for the poor, growth in G.N.P. is still a good thing, but for the rich it is probably a bad thing. Growth in population, however, is bad for both.

Daly's definition of a 'steady-state' economy is one in which a constant stock of physical wealth (described as 'capital') and of people (population) is maintained. In other words, population must be kept constant and in line with an appropriate constancy in physical demands upon the earth. These stocks must be maintained by a rate of 'inflow' equal to the rate of 'outflow', i.e. birth and production must be balanced against death and consumption to maintain a steady state in the economy as a whole.

Daly calls this process 'throughput', and points out that even this definition of a steady state is not of itself sufficient to ensure a sustainable economy. His definition of a steady state thus includes consideration of *the rate of 'throughput'*:

For a number of reasons we specify that the rate of throughput should be as low as possible.

A low birth rate and an equally low death rate, which would represent a low rate of throughput for the population, means a high life-expectancy. For the stock of wealth, low rate of throughput necessitates greater durability of goods, and results in lowered consumption of resources and less pollution. The limits regarding what rates of depletion and pollution are tolerable must be supplied by the ecology.

The above criteria for a healthy and sustainable economy make fewer demands upon our environmental resources, but clearly demand a high degree of moral responsibility and cooperation, and Daly notes that there is simply no substitute for moral resources. Moreover, such are the dangers inherent in the very high degree of technical expertise of twentieth century man that we have no alternative but to rely on moral restraints. While this is undoubtedly true in principle, I can see little evidence that moral restraints are sufficient to contain the frenetic pursuit of wealth in which so many people in the world are involved. This is one more reason why we would be wise to seek out quite other paths to sanity.

Daly suggests that if physical stocks are to be kept constant, economic growth must be in the form of nonphysical goods - that is to say there must be a growth in service-oriented activity and in leisure, a scenario quite foreign to historical practice, and there will be significant social connotations to this change.

A fascinating sidelight on this aspect of economic growth is that as a result of the improvements in technology, and the subsequent increase in production of goods per unit of time, time has become exceedingly expensive, and fewer and fewer activities are 'worth the time'. We become what Daly calls 'goods-rich and time-poor' - or, as Staffan Linder3 puts it, we have created 'the harried leisure class'. Thus time-intensive activities, such as friendships, care of the aged and children, meditation and reflection, are sacrificed in favour of commodity-intensive activities, which is exactly the opposite of what Daly sees as the requirement for an ecologically sound economy.

Daly's remedy is to keep physical stocks constant and use technology to increase leisure. This idea is of course not new. Daly quotes Bertrand Russell's now famous essay 'In Praise of Idleness' in support of the latter's proposal, made in 1930, that economic growth should be directed toward leisure growth rather than commodity growth. Russell's delightful essay discusses the unemployment question in terms that must bring a smile to the sternest lips:

> Suppose that, at a given moment, a certain number of people are engaged in the manufacture of pins. They make as manypins as the world needs, working, (say) eight hours a day. Someone makes an invention by which the same number of men can make twice as many pins as before. But the world does not need twice as many pins. Pins are already so cheap that hardly any more will be bought at a lower price. In a sensible world, everybody concerned in the manufacture of pins would take to working four hours instead of eight, and everything else would go on as before. But in the actual world this would be thought demoralizing. The men still work eight hours, there are too many pins, some employers go bankrupt, and half the men previously concerned in making pins are thrown out of work. There is, in the end, just as much leisure as on the other plan, but half the men are totally idle, while half are still overworked. In this way it is ensured that the unavoidable leisure shall cause misery all round instead of being a universal source of happiness. Can anything more insane be imagined?

The curtailment of commodity growth inevitably impinges on the very foundations of economic thinking, and in particular on the vexed question of the distribution of wealth. Daly notes that the Marxist doctrine of the classless society is based on the premise of complete abundance, which is manifestly faulty in the light of current knowledge, since the ecosystem provides a set of physical constraints to which all economic systems must conform. 'The facility with which an economic system can adapt to these constraints is a major, if neglected, criterion for comparing economic systems.'Needless to say, the ruthless exploitative practices of capitalist production, and especially of big capital in pursuit of profit, are equally far from working within these restraints.

'Why', asks Daly, 'do people produce junk and cajole other people into buying it? Not out of any innate love for junk or hatred of the environment but simply to earn an income. If, with the prevailing distribution of wealth, income and power, production governed by the profit motive results in the output of great amounts of noxious junk, then something is wrong with the distribution of wealth and power, the profit motive, or both. We need some principle of income distribution independent of, and supplementary to the income-through-jobs link.'

It is beyond the scope of this book or the competence of its author to discuss current economic thinking on these vital but extremely complex problems. I can only add that I believe we need changes even more radical than Daly suggests, involving the distribution of goods, services and information transcending any notion of income-through-jobs. I hope it will become clear, as my book proceeds, how this might be done.

Hazel Henderson

Two strong arguments stand out in my mind from a reading of Henderson. Firstly, that economic thinking has strayed too far from the ecological base that constitutes our primary economic system, and that we are in consequence being overtaken by larger global forces which threaten our survival. Secondly, that by far the greater part of economic activity has been left out altogether from the thinking of economists.

Henderson argues fiercely that failure to recognize the source of all material wealth, and our tendency to regard the resources of the Earth as infinite, are the main contributing factors to economic instability. In trying to deal with the 'strange new diseases' of structural inflation and structural unemployment, economists are 'continually rationalizing away the real reason for them'. Those reasons, in Henderson's view, are better understood outside the discipline of economics. 'They are rooted in the way we use resources and raw materials, and they are rooted in the particular capital-and-energy-intensive type of technology that we have developed.'

It was, of course, inevitable that this realization would intrude into economic thinking sooner or later. Indeed, had it been sooner instead of late, we'd be a lot better off. There's a growing chorus of voices from the scientific world warning us that unless we take immediate and effective action, we face the actual destruction of mankind.

Ironically, it is economic issues that tend to put the brakes on corrective action. We measure the monetary cost of depletion or pollution controls, and, nine times out of ten, decide that we cannot afford them. The result is a frightening trend toward irreversible destruction of the Earth's capacity to renew herself. That Henderson adds a powerful voice to opponents of this trend can only be welcomed, but it is her second theme that impinges most closely on my own work. It concerns what she is pleased to call 'the hypnotism of money', as a result of which, in her view, all schools of economic thought look only at those sections of production and consumption in their countries that are monetized and involve cash transactions.

> This colossal error of equating the monetized half of most industrial economics with the whole system of production, consumption and

> maintenance is common to all branches of economics and accounts for its one-dimensional, linear, partial view, as opposed to the wider realities of seeing economics whole.

This view crops up in the work of many thinking people today. We become increasingly aware that the people streaming into the cities in search of largely non-existent employment were once self-reliant; that they belonged at one time to tightly knit village communities whose wealth lay in their women, their cattle and sheep, and their crops; to nomadic tribes with their flocks and herds, or to the primitive hunter-gatherers of many lands, whose wealth lay in rivers teeming with fish or plains alive with grazing antelope.

Henderson points out that the economies of these self-sufficient peoples rested on reciprocity, barter, mutual help and the re-distribution of surpluses according to ancient customs and traditions. It all sounds idyllic, provided you close your thoughts to inter-tribal warfare, the theft of cattle and women, the burning of villages, periodic famine, and endemic disease. Nevertheless, Henderson's point is valid enough. Industrial development has undermined the very basis of rural self-sufficiency, leaving people adrift like so much flotsam on the edges of a society which has no room for them, and in which, having forfeited the culture and way of life of their forefathers, they are ill-equipped to participate.

> One of the aspects of the crisis of industrial development is that it begins to suck all such informal, use-value production and consumption into the monetized economies, drawing populations into the cities, denuding agricultural areas, dissolving the cultural glue of village life and reciprocal community systems of food-sharing, care of the young and the elderly, and folk medicine, and destroying inherited cultural wisdom.

Henderson points out that as industrialization and monetization colonize more and more of the informal, use-value production and consumption, limits to this process are reached. These limitations finally show up in attempts to 'monetize' environmental resources, and to place cash value on air, water, open space and even human life, since they undermine the living, caring relationships by which people in all societies have traditionally rendered each other vital services entirely free of charge.

This insistence on using money as the only variable to measure the efficiency of production and distribution thus emerges in Henderson's work as one of the main errors in all current schools of economic thought.

She goes so far as to observe that there is perhaps no better way of summing up what economists call 'inflation' than to define inflation, from beyond economics, as simply all the factors and variables that economists have left out of their models.

One remedy involves correcting the long-standing conceptual error of equating our society's total socio-economic productivity with that portion of it based on competitive, market-based cash transactions and the flows of money they generate. Another, now widely under discussion, is the recognition of energy as an important variable for measuring economic activities, a concept that introduces the entropy factor into economics. Henderson points out that in many of today's most highly industrialized nations, the dissipation of energy into unproductive activities such as mediating conflicts, controlling crime, or managing large bureaucracies, has reached such proportions that their cost makes up an ever-increasing proportion of the Gross National Product thus driving inflation to ever greater heights.

Emphasis is laid on the importance of the informal sectors of the world's economies, which, contrary to general belief, constitute a major part of productive activity, and on the merits of the small company in affording opportunities for personal development and job satisfaction not easily available to the majority of the labour force in giant corporations. As Henderson observes in *The Politics of the Solar Age*.

> All this is not to say that large corporations will fade away. They will not. We will still need to pour steel and aluminum and maintain telephone systems and electric grids. But some of these systems may already have reached an optimal size, and the growth of newer sectors of the economy may better satisfy new needs in wholly new ways for which the old corporations and their existing technological configurations may be quite unsuited.

I doubt that many economists will like Henderson's analyses, calling in doubt the very foundations of their discipline, but such is the condition of the world economy, and, in particular, of the world monetary system, that it is difficult to refute her arguments.

In so far as they impinge on my own work, perhaps the most useful of her insights is the extent to which production, consumption and maintenance proceed outside the monetized economies. An extension of these principles to a non-monetized, information-based economy would seem a natural progression.

CHAPTER TWENTY SIX

6.9 The Systemic Approach

Summary: The systemic view is an extension of the principles of ecology to all aspects of human endeavour, including the psychological and spiritual. Its application in the field of economics is particularly urgent.

The systemic view could be thought of as an extension of the principles of ecology to include the psychological and spiritual aspects of human life, a view entirely in keeping with my own approach.

The most comprehensive presentation of this view I have read is Fritjof Capra's excellent book *The Turning Point*,[1] in which he suggests that the new concepts in physics have brought about a profound change in our world view; that the mechanistic conception of Descartes and Newton have given way to a holistic and ecological view; and that this view is found to be similar to the views of mystics of all ages and traditions.

Capra makes the point that in their struggle to grasp this reality, scientists became painfully aware that their basic concepts, their language and their whole way of thinking were inadequate to describing atomic phenomena, a situation which created not only intellectual malaise, but intense emotional distress amounting to what might be called an existential crisis. The basic thesis of Capra's book is that the social turmoil in which society finds itself today is a facet of this same crisis, deriving from the fact that we are trying to apply to our problems the concepts of an outdated world view. What ails us is an outdated perception of reality. The final emergence of scientists from this state of mind has enriched them with profound insights into the nature of matter and its relation to the human mind. In Capra's words:

> We live today in a globally interconnected world, in which biological, psychological, social and environmental phenomena are all interdependent. To describe this world appropriately we need an ecological perspective which the Cartesian world view does not offer.

In the brief exploratory look at aspects of current thinking so far described, we have been seeking what might be called 'pointers to survival' for mankind. Partial solutions have emerged from a number of related disciplines, and Capra's work is in one sense an attempt to synthesize their differing perspectives. His book deals with the emerging systems-view of life in its application to mind, consciousness and evolution; with the corresponding holistic approach to health and healing; with the integration of Western and Eastern approaches to psychology and psychotherapy; and with a new conceptual framework for economics and technology. He offers an overall ecological perspective 'which is spiritual in its ultimate nature and will lead to profound changes in our social and political structures.'

In so far as it impinges on my own work, the most significant parts of Capra's work relate to the systems-view of life, to ecological perspectives, and to economics, and though they cannot be dissociated one from another, I want to look in particular at the last of these. We shall see that in Capra's view as in mine, the systems-view of life is an appropriate basis not only for the behavioural and life sciences, but also for the social sciences, and especially for economics.

> The application of systems concepts to describe economic processes and activities is particularly urgent, because virtuallyall our current economic problems are systemic problems that can no longer be understood via Cartesian science.

Like a number of other authors already mentioned, Capra stresses that most economists tend to dissociate the economy from the ecological fabric in which it is embedded, and that the highly imbalanced set of values that dominate our culture has led to an over-emphasis on hard technology, wasteful consumption, and rapid exploitation of natural resources, all motivated by the persistent obsession with growth. He suggests that the systems approach to economics will make it possible to bring some order into the present conceptual chaos by giving economists the urgently needed ecological perspective.

> According to the systems view, the economy is a living system composed of human beings and social organizations in continual interaction with one another and with the surrounding ecosystem on which our lives depend ... The recognition of the non-linear nature of all systems dynamics is the very essence of ecological awareness.

Capra suggests that this non-linear interconnectedness of living systems immediately invites the application of two important rules for the management of social and economic systems. Firstly, there is an optimal size for every structure, organization, and institution. It follows that to maximize any single variable - profit, efficiency, or G.N.P. for example - will inevitably destroy the larger system. Secondly, the more an economy is based on the continual recycling of its natural resources, the more it is in harmony with the surrounding environment.

A healthy economy will thus depend on keeping in balance the dynamic interplay of fluctuating and complementary tendencies. In order to maintain a healthy economic system we need to cherish the innate flexibility of our natural environment and to cultivate a corresponding flexibility in our social institutions.

> What we need then, is a new 'paradigm'- a new vision of reality; a fundamental change in our thoughts, perceptions and values. The beginnings of this change ... are already visible in all fields and are likely to dominate the present decade.

Capra notes that the sixties and seventies have generated a whole series of social movements that all seem to go in the same direction, emphasizing different aspects of the new vision of reality. So far, most of these movements still operate separately and have not yet recognized how their intentions interrelate. So the purpose of his book is:

> to provide a coherent conceptual framework that will help them recognize the communality of their aims ... Once this happens, we can expect the various movements to flow together and form a powerful force for social change. The gravity and global extent of our current crisis indicate that this change is likely to result in a transformation of unprecedented dimensions, a turning point for the planet as a whole.

I can only endorse this view.

BOOK II : Part Seven

Gaia and the Microchip

Principles of Living Organisms Relevant to Human Society

CHAPTER TWENTY SEVEN

7.1 Living Principles

Go out into the darkness and place your hand in the hand of God. That shall be to you better than light and safer than a known way.

Louisa Mary Haskin in '*The Small Woman*'

Principles that emerge so far from a study of multicellular living systems are:

- They are open systems, in continuous interaction with their environment.

- They are self-governed.

- They are based on each fulfilling his intrinsic capabilities, right down to the molecular and atomic level.

- There is an element of choice.

- Choice is always in the interests of the system.

- No unit is expected to do or be anything other than itself.

- Self-interest finds its best expression in the good of the whole.

- The whole is best served by maximum realization of its parts.

- Coherence results from the relationships between two or more entities, in ever increasing complexity.

- This leads, in the more complex organisms, to specialization, i.e. organisms develop organs.

- The interaction of even two cells requires that information should be exchanged between them. The more complex the organism, therefore, the more encompassing its information systems.

- Organisms and environment are one and the same system, neither separate nor separable one from the other.

- Information exchange involves cybernetic principles of feedback. This leads to automation of certain functions.

- Primary relationships are contiguous. So groupings proceed from local to regional to general - not the other way round.

- There is no voting; no power blocks. The highest authority, the brain, emerges from interaction of individual cells fed by, and in turn feeding, all cells of the body. No single cell, while it lives, is exempt from participation or can survive in isolation.

- Information proceeds from, and is immediately available to, all parts of the body.

- Variations in form, in function, and in association are infinite, so one must not look for uniformity.

- There is room for every kind of association, however bizarre. We must not presume to dictate what people are or should be.

- It is this diversity that gives life its interest, its beauty, and its survival capacity.

- Individual association is brought about by instinctive and by functional affinity, i.e. by mutually beneficial possibilities and their recognition through an exchange of information.

- Functional coherence does not depend on utopian transformation, but on the ability of constituent parts to form relationships, and to engender thereby something of value in some part of the system.

- The well-being of any system, large or small, is dependent on the well-being of all its constituent parts.

- The drive to self-preservation is too fierce to allow of boredom or the pointless destruction of the environment.

- Natural systems are extremely fluid and adaptable. If they are not so, they die.

- There are, at least in complex organisms, mechanisms for storage of surplus foodstuffs, which can be released in time of need.

- Trigger mechanisms set in train a host of interlinked processes.

- Certain processes appear to be catalytic.

- The pattern appears to remain relatively stable while individual parts come and go. Intangibles are therefore the dominant factor in maintenance of structure and function.

- Natural systems are always multi-dimensional webs - never linear.

- In a multicellular organism there is no such thing as power for power's sake. (In the wider relationships of nature, power is always linked to an imperative for survival, such as territory, sexual predominance, and so forth, but in an articulated living system power becomes irrelevant and is non-existent.)

- Every natural system appears to have at least three phases: intake, usage and waste disposal.

- Every phase of one natural system inevitably influences, maintains, enhances or inhibits the life processes of another. If the system is to have long survival value, its processes and those of its environment must be mutually supportive in this regard. Directly or indirectly they must contribute to the healthy ecology of the entire planet.

- The most important specialization in any articulated system is the development of brain cells or neurons, whose capacity for forming links and circuits provides the means for coherence and self-governance of the whole system.

CHAPTER TWENTY EIGHT

7.2 The Dawn of Economics

Summary: Money and the market economy developed prior to the advent of computers, in an age when information as we know it today did not exist, and when the concept of a living Earth was yet to be born.

With the aid of computers, money as an expression of value in exchange can now give way to cybernetic principles in the regulation of economic activity.

I am up to my ears now in a morass of my own making, plunging into a pathless jungle of ideas concerning man's relationship with man, and with the practicalities of his day to day existence. I have to hold on, for all I'm worth, to the central idea - that the Earth is alive - so I don't get lost. Somewhere in the jungle are strands which can be drawn to together to answer the old and baffling question that so many people tend to ask of me:

'So what? What difference does it make if Earth is alive? Can it help?'

And because life today is increasingly and frighteningly dominated by money or the lack of it, 'so what' boils down to this:

> How, in the light of a living Earth, can the economic structure of society be modified or recast to give everyone at least a sporting chance, and ennoble each man's life?

I'd caught a glimpse of how, with the help of the microchip, this could come about, but when it came down to practicalities, I just didn't know enough. I plunged in headlong and began to drown, for now I was well and truly out of my depth. I'd often bogged down before in reading beyond my powers, and when this had happened, had ruefully abandoned the books. This time I had to plod stubbornly on, for whether we liked it or not, economic restraints appeared to control our lives.

We speak of freedoms and justice and human rights, but when it comes to action our hands are tied. Economic concerns dictate what we do and are. The way we produce; the way we buy and sell; the wages we earn; the power of capital, unions or rents; whether the interest rates are low or high; the price of gold; and how much we must pay for a loaf of bread - these are the things that dictate our daily lives. I had to get a perspective on how it all worked.

None of us has anything useful to say from our own exclusive knowledge. We borrow and assimilate and build on other minds, whose interlocking thought has fashioned us. If we are very lucky, we add a small new insight here or there. So I knew it wouldn't be fruitful trying to start from scratch with a living Earth, however clear the principles might be. Economic systems, like everything else, were rooted in history. They belonged specifically in time and place; to the way people lived and worked in that time, and in that place; to the way they adapted to circumstance; and to the way they thought. Our own time and place wouldn't be different from this. The 'now' of our times would be rooted in the past. I had to get at it.

I needed a lot of help with this, and to my great good fortune, came upon, in the midst of my confusion and ignorance, a Pelican Original first published in 1967 called *A History of Economic Thought* [1] by William J. Barber.

This excellent little book gave me a framework within which to fit previous desultory reading, and an historical perspective against which to assess my own somewhat unconventional thinking. I acknowledge it gratefully.

In the words of Professor Barber, then:

> Each of the intellectual systems ... examined - i.e. those of classical, Marxian, neoclassical, and Keynesian thought - yield different insights into the nature of the economic universe and into the ways in which men can most effectively come to grips with it.[2]

To my surprise, the ideas that wound their way in different forms through all these early discussions were still the subject of debate today, as were the problems of international trade and the balance of surpluses.

Early economic theory sprang from an age nearly three hundred years ago when mercantile fleets were scouring the world for trade, and precious metals were the chief means of exchange. Power at home and abroad sprang from silver and gold.

> [Countries] lacking sizeable and exploitable deposits of gold or silver were obliged to draw on foreign supplies. A favourable balance in the international accounts was thus a precondition for substantial enlargements of the money supply called for by a prospering and expanding economy.[3]

After nearly three hundred years of international rivalry, often backed up by war and the decimation of people's personal lives this description of seventeenth and early eighteenth century economics has a poignant and sadly familiar ring.

Formal economic systems, however, seem to date from 1776, the year in which Adam Smith published his now-famous book, *An Inquiry into the Nature and Causes of the Wealth of Nations*.[4]

Smith was concerned first and foremost with economic growth over long time periods. At a time when, as now, large numbers of people lived below the breadline, when industrialism was still in its infancy, when commodities were scarce, and when resources of the earth appeared to be infinite, this was of course entirely appropriate. Smith could hardly have been expected to foresee a time when enlightened economic thinking would focus on the restraints of a finite earth. The limitations to growth built into living systems were not yet visible to Smith's generation of men.

As he viewed the matter, workers engaged in certain occupations were more likely to promote the advancement of output in the future than those employed in others. He developed the point by asserting that the 'productive' employments must meet two tests:

1. that they led to the production of tangible objects, a condition prerequisite to accumulation; and

2. that they gave rise to a 'surplus' that could be made available for future re-investment.

For most practical purposes he equated the 'productive' employments with those in which labour worked with capital.[5]

In Smith's view, then, only activity which led to the growth of tangible material objects could promote the national wealth whereby the general standard of living of the population could be enhanced. Virtually all 'service' activities were excluded. Barber notes that Smith's division of human labour into 'productive' and 'un-productive' categories was not intended as a value judgment on human worth, nor on the importance of 'unproductive' professional activity, but with the capacity of 'productive' activity to promote long-term economic change. 'Smith', says Barber, 'drew a sharp line of demarcation between 'value in use' and 'value in exchange'. He found only the latter economically interesting.'[6]

In view of the tremendous influence of Adam Smith's work on later economic thinkers, it is easy to see how this material aspect of 'wealth' became entrenched in the assessment of Gross National Product on which we operate today, and to which Hazel Henderson takes such exception.[7]

While modern economists would include in the national income all service activities to which a monetary price could be affixed, the countless unpaid services which hold society together remain outside their calculation of national income.

Listening each morning to the economic report on our daily radio news, I marvel that all this abstract analysis of gold price and shares and trade deficits and currencies can have anything to do with peoples' daily lives. It is all so far removed from personal needs. It feels quite alien. It could be coming from somewhere out in space. Yet in its forecasts and fluctuations are mirrored for most of our citizens whether they work or stand idle, whether they eat or do not eat, whether they live or die.

I need hardly say that I am in no position to express an opinion on the merits or demerits of economic analysis, past or present, still less to develop a new system. But having in mind the possibility of an economic system in which monetary values might play a diminishing role, this kind of discussion no longer appears to me as central to human well-being. It belongs to an age in which the market provided a link by which to co-ordinate economic activity; when information as we know it today did not exist; and when the concept of a living Earth was yet to be born. What I hope to do is to point the way to an organic approach which might lend a new dimension to the thinking of people better equipped than myself to explore its usefulness.

Money has had at least three roles in human society:

- As a means of exchange.

- As a means of capital accumulation by which the means of production could be increased.

- As a bearer of information concerning man's needs and wants.

We are still involved in all the same basic problems: with the means of exchange involving value and price; with capital accumulation and productivity; and with the market as an indicator of need. It is the last of these that is crucial to our argument.

The point is succinctly made by Professor F.A. Hayek in a recent essay entitled 'The Moral Imperative', published in 1986 by the Institute of Economic Affairs in a book of essays entitled *The Unfinished Agenda*.[8]

In 1936, the year in which (entirely coincidentally) John Maynard

> Keynes published *The General Theory*, I suddenly saw ... that my previous work in different branches of economics had a common root. This insight was that the price system was really an instrument which enabled millions of people to adjust their efforts to events, demands and conditions, of which they had no concrete, direct knowledge, and that the whole co-ordination of the world economy was due to certain practices and usages which had grown up unconsciously.

Hayek explains how he gradually found that the basic function of economics was to explain how human activity adapted itself to data about which it had no information; that the whole economic order rested on the fact that by using prices as a guide, or as signals, we were led to serve the demands and enlist the powers and capacities of people of whom we knew nothing. He observes further that because we relied on a system which we had never understood, and which we had never designed, we had been able to produce the wealth to sustain an enormous increase in the world's population, and begin to realize our new ambition of distributing the wealth more justly.

> Basically, the insight that prices were signals bringing about the unforeseen co-ordination of the efforts of thousands of individuals was in a sense the modern cybernetics theory, and it became the leading idea behind my work.[9]

This intuition is surely a step in the right direction. But we now have at our command an infinitely more competent instrument than price for co-ordinating human activity, one which can moreover be used to bring about an equally spontaneous but far wider degree of coherence in society at large. We must therefore look at the advantages and limitations of price as a regulating factor, before going on to consider the electronic tools available to us for doing the same job. For the sake of clarity, only the idealized concept of free market transactions need be considered at this point.

Price is the money expression of value in exchange, and is determined in the market. People buy and people sell. If they need or want certain goods, demand soars. If prices are too high, the sales fall off. Either way information has been exchanged concerning people's wants or needs, by means of the prices they are prepared to pay. In this way prices become the co-ordinating signals Professor Hayek talks about so convincingly.

But if prices are signals, or bearers of co-ordinating information, how much information are they equipped to convey? In Stephen Bodington's words:

> The idealized concept of the market system sees it as determining through competition the level of prices at which the seller gives and the buyer receives equal economic values.[10]

This definition sounds fair enough, provided there is consensus concerning the meaning of value, which there is not. Barber suggests that a mid-twentieth century economist, asked to state the 'value' of a particular commodity, would normally proceed by trying to establish the price the market was prepared to pay for it.[11] Writers in the classical tradition, on the other hand, viewed the concept of 'value' as independent of the market's whims.

> Nominal (or market) prices might fluctuate, but value remained constant and invariant.[12]

Adam Smith based the 'value' of a commodity on the quantity of labour that would be required for its production, using as his standard a hypothetical 'early and rude' society such as might be found in a nation of hunters:

> If ... it usually costs twice the labour to kill a beaver which it does to kill a deer, one beaver should naturally exchange for or be worth two deer.[13]

When it became evident that the appropriation of private property and the accumulation of capital made this interpretation invalid, Smith abandoned the labour 'content' aspect of value, and suggested that 'command' over labour was the appropriate measure of value.

Later thinkers, from Ricardo to Karl Marx, refined and developed the concept of labour input in the assessment of value, each in turn wrestling with the particular aspect of industrial activity most pertinent to the time in which he lived, and the sophistication of industry in a rapidly expanding economy. Ricardo, for instance, viewed a machine as 'embodied or accumulated labour, the productive powers of which would be transferred to current output over the course of its life,'[14] thus including capital in the argument, while maintaining labour input as the appropriate measure of value.

Ricardo became, however, increasingly aware that the accumulation of capital introduced too many variables into the equation, and in his analysis, the labour interpretation of value could not be fully sustained. It was to reappear with banners flying in the work of Karl Marx (1818-83).

We shall see during the course of my own work that the concept of value attains a totally different meaning in an economy based on organic principles.

Meanwhile John Stuart Mill had reopened the argument on 'productive' and 'unproductive' labour.

> In his view the argument that labour was productive only when it produced material objects called for a re-examination.[15]

Mill suggested that the transmission of skills should be included as productive labour, 'provided that an increase of material products'[16] was its ultimate consequence. This insight was later to be narrowed down still further by Marx, who insisted that labour-time involved in commodity production only created value where such inputs were 'socially necessary'. This qualification was to have far-reaching consequences in Soviet bloc economics, for it pre-supposed some kind of authority empowered to decide which of a man's activities fell within socially necessary limits.

Marx added a further restriction to his analysis of value:

> Production for exchange, he asserted, was a prerequisite of value. Pre-capitalistic arrangements could produce outputs, but by Marxian definitions, they could produce neither commodities nor value.[17]

Given these conditions, Marx was convinced that exchange value (or relative prices) was controlled by the labour inputs required in the production of commodities. Under capitalism, exchange ratios between commodities could be expressed in money terms only 'because all commodities, as values, are realized human labour, and therefore commensurable.'[18] Their values could therefore be measured by a common commodity, and the latter expressed as money.

It would appear, then, that throughout the pre-twentieth century history of economic analysis, the concept of price as a *bearer of information* was overshadowed by its role as a *measure of value*. Today the focus of interest has shifted radically, and while market concepts are still a matter of debate, 'price', and therefore money itself, are widely recognized as 'information in motion'.

Writing in 1985 in his book *The Silicon Idol*,[19] Michael Shallis has this to say about money:

> Money, that is coinage, has traditional and symbolic significance. On the most mundane level it symbolizes a person's ability and choice to use his or her labour in exchange for a token.

Shallis quotes Joseph Weizenbaum as writing that people once traded their labour directly for goods. Then money became an abstract quantification of human labour. Cheques and other financial instruments followed, becoming abstractions for money. Now we are moving towards the so-called 'cashless' society in which electrons racing around computers become abstractions for financial instruments.

> An observer from another planet will see people labouring in order to optimize the paths of electron streams flowing on their behalf in computers unseen and incomprehensible.[20]

While money as coinage may have been superseded by information, it is important to note that price, as determined by the market, is still at the centre of economic debate. It remains for the concept of a living Earth to introduce a new means of co-ordinating human activity - a means which, while maximizing individual worth, will make no attempt to assess the *relative* values of human input. At this point information itself will become the means of exchange.

CHAPTER TWENTY NINE

7.3 Oblique Barter - The Economics of Indirect Benefit

Summary: Electronic technology can achieve for global society an intimate, face to face integration in which self-interest and communal benefit are perceived to be synonymous.

When you look at Economics in the light of a living Earth, the most significant thing to emerge is that, in nature, *benefits from one organism to another are not usually direct*. Each pursues his own activity, drawing on sunlight, water, air, soil or other organisms, but the net effect is enhancement of the environment for all of them, *because each helps to create the environment of all the others*. No-one demands or expects extraneous payment for services rendered. Nor does any organism pause to wonder whether the benefit he bestows is equal in value to the benefit he receives. Benefit springs from the random interweaving of all nature's activities.

This observation gave me to wonder whether human society would lend itself to a similar mode of organization, where it no longer mattered whether what I did for you was worth less or more (in monetary terms) than what you did for me; where what I received from you did not necessarily have to be paid for direct by you, nor vice versa; and where you and I in turn would both receive benefit indirectly from the interrelated activities of everyone else.

We in the West are so steeped in the monetary tradition, and are daily made so cogently aware of its dominance over our lives, that it is difficult to conceive of such a world. Western economy is dominated by price, by the 'market', by supply and demand, and by the unchallenged assumption that what one receives must be paid for directly, either from man to man, man to firm, firm to firm, or by taxation for some general social good. But in nature, benefits from one organism to another are by no means always directly reciprocal, in spite of a number of striking symbiotic relationships in both animals and plants. I found myself wondering whether we had perhaps outgrown the current economic mode; whether perhaps, in the human field too, *goods and services need not necessarily be paid for directly to the person or persons from whom they were received.*

Sometimes it is natural and convenient to exchange, say, a bag of potatoes for a bushel of wheat, or the service of a doctor for a kilo of fish. But this is not usually so. More often we use a means of exchange to enable us to pay the doctor or purchase the wheat. The crucial point is this: we still find it necessary to involve ourselves in a system of prices and values, and to render payment directly of any such sum as the market demands.

We have of course various quite sophisticated ways of doing this, but that is not the point at issue here. What is important in the context of a living Earth is that neither nature nor any living organism usually works this way. When it comes to the internal arrangements of physiological process, the absurdities of value assessment, of price, of capital accumulation or the acquisition of power become even more apparent. Intercellular coherence just couldn't work this way.

So what I am suggesting is this: *that we move away from the value assessment of goods and services on which our monetary economy is based*, towards a society in which personal and general benefit are achieved by the spontaneous activity of every human being in the sphere in which his happiest interest lies, and in which, because *he is thus motivated, he is at maximum productivity.*

I am suggesting that these activities should not be paid for in the sense in which we think of payment today - i.e. in wages or profits or rents or interest - but that they be co-ordinated by means of the computers and electronic devices already at our command; and that the net result will be a flow of goods and services adequate and satisfying to the body of a healthy society, as in other living organisms. I have given the name of '*oblique barter*' to this process to signify its indirect nature. However 'network barter' would be equally descriptive, in recognition of the extraordinarily complex nature of modern industrial society.

There are of course numerous direct and mutually beneficial arrangements in nature. Some of these are well known, as that the bee in his pursuit of pollen, fertilizes the flower. Others less well known are equally ingenious and sometimes downright bizarre. But the broad spectrum of nature's operations contains a principle we have yet to understand:

Provided you contribute something of value to *someone*, or to the matrix as a whole, *it does not matter to whom or through whom you offer it.* You do not have to recompense directly the person or persons who benefited you. His need is sufficient to warrant service from one or more of his fellows, and your capacity and will to render it makes you eligible in like manner for the fulfillment of your need. This is a little different from the famous socialist ideal:

> From every man according to his capacity. To every man according to his need . . .

because this latter involved the intervention of the state. Worthy though the ideal undoubtedly was, in practical politics it led to a bureaucracy which decided both a man's field of contribution and his allowed quota of need, thus negating his freedom of choice, inhibiting his initiative and reducing his output. In nature there is no such bureaucracy. The very diversity of species, climate and condition forbids it, ensuring a self-chosen niche for the activity of every imaginable creature and kind. *Because of this diversity, the indirect exchange of benefit is viable.*

Applying these ideas to humankind we remember that the practice of indirect reciprocal benefit is very old. Primitive man and tribal societies have always worked this way. The hunter went out after living prey. The women gathered berries and roots, and tanned the hides. Children gathered firewood, and, in pastoral society, herded the sheep or goats. Labour was divided. So was food. Each did what he was equipped to do. Even the ancient women of the tribe had specialized tasks, like keeping alive the precious spark of fire. But while there was food to eat, everyone had a share. Everyone knew his survival depended on the survival of all. The well-being of the tribe sprang from a network of inter-supportive activities.

So far so good. But all the members of a tribe knew each other. There was face-to-face intimacy between them - an intimacy of association which became impossible when populations grew and prospered and spread their millions across the land.

Hayek makes this point very well in the essay cited above (p.126). He observes that agreement about a common purpose between a group of known people is clearly an idea that cannot be applied to a large society which includes people who do not know one another. The modern society and the modern economy have evolved in recognition of this limitation. A face-to-face intimacy is simply not applicable to large groups.

> The essential basis of the development of modern civilization is to allow people to pursue their own ends on the basis of their own knowledge, and not be bound by the aims of other people.[1]

One may note in passing that this kind of self-interest, *without commitment to a common purpose* has led to the lethal competitive and cut-throat situation in which we find ourselves today. While it was perhaps inevitable in the period of history in which it occurred, it is no longer either necessary or wise.

Hayek sees 'price' as the instrument of modern society which has enabled people to proceed with this kind of economic autonomy.

> In order ... to enable [people] to adapt themselves to a structure which they do not know (and the determinants of which they do not know), we have to allow the spontaneous mechanisms of the market to tell them what they ought to do.[2]

Hayek observes that the same dilemma applies to the basic desire of socialism for distribution according to principles of justice:

> If prices are to serve as an effective guide to what people ought to do ... [you] must allow prices to be determined so as to tell people where they can make the best contribution to the rest of society - and unfortunately the capacity of making good contributions to one's fellows is not distributed according to any principles of justice.

While respecting the reach of this author's insight, it is with a sense of sadness one has to add the coda that large numbers of people are not very interested in making the best contribution they may to the rest of society. They are, however, very interested indeed in making the best contribution they may to their own well-being, so we have to build on this. Self preservation is, after all, a healthy biological principle, however sad we may feel that we are not always able to transcend it.

Happily, the process has come full circle in the present decade. Inherent in the electronic revolution is the capacity to retrieve the person-to-person intimacy once characteristic of the tribe, and to understand full well what we are about. The 'global village' is about to come of age.

Bearing in mind the mind-boggling capacity we now have to integrate people's desires and capacities, it is probably safe to say that there is no aspect of self-interest, and no human skill that cannot be usefully applied in the new order once the new paradigms have caught people's imagination.

The one overriding question, then, that we must ask ourselves when pondering these matters is this:

Is it in line with the accommodations we find in living systems?

If it is, we are on the right track. If it is not, the idea will be sterile, leading us nowhere.

CHAPTER THIRTY

7.4 Requirements for a Friendly Society

Summary: Ethical signposts towards an information-based economy.

- If an information-based society is to rely on computers for integration and regulation of exchange in goods and services, then all information concerning individual needs, preferences, enthusiasms or capacities must be fed into the system by the individual himself, or by his representative, i.e. the system must be self-regulatory at grass-roots level, and wholly spontaneous.

- The system must give all human beings equal access and opportunity, *provided they bestir themselves to contribute to the full,* using whatever skills and talents they possess.

- It must offer infinite opportunity for mental expansion and self-improvement.

- Participation must be optional for maximum freedom. It must not be imposed. Growth in usage must come naturally, and must not be confined within an ideology nor channelled toward one.

- The system must not concentrate power.

- It must recognize and encourage the precious ethical intangibles, and enhance human vision.

- It must be exploratory, and must evolve. It must not be prescribed.

- It must remain fluid, open and unpredictable.

- It must carry man ever beyond himself.

- It must encourage respect for other living things, and reverence for life.

- It must feed the human spirit, enhance beauty and kindliness and incline the soul to God.

In order to embody these principles in socio-economic and political values, we have to find more efficient ways of integrating human activity than those at present in use. I have suggested that this can only be done if we start by recognition of the Earth as a living system in its own right, and then go on to use our human inventive and technological expertise in the service of this system.

I have suggested that it is part of the evolutionary process that we have developed these skills at exactly the moment when we need them most, and that their proper and appropriate use will help to trigger the next evolutionary leap.

I have suggested further that as in any living organism, automation of certain functions is a prerequisite for the development of higher cerebral activity, and that the evolution of man, and of planet Earth, proceeding together, are on the critical threshold of this automation.

In the economic field, I am suggesting that the new technologies permit us to replace the 'models', 'growth curves', and other sophisticated tools of economic analysis by a moment-to-moment factual regulation of utilities, based on people's own assessment of what they desire to do, or what they need; and that though, in the system envisaged, the overall indices of these social networks may not be readily accessible by any bureaucratic body, their net effect will be to produce what people actually want, and to reflect the actual situation of any society at any given moment in time.

Here it is necessary to stress again that the idea is not to supply some governmental body with information concerning each individual, but to enable each individual or group to contact each or any other individual or group at will, proceeding outwards from his own local computers and data banks in ever widening circles and networks. Economic activity will thus proceed from grass roots level, not the other way round, and will have characteristics akin to the vast network of capillaries that are in contact with every cell of a living body, and through which all exchange between tissues and the blood takes place; or with the yet more finely articulated structure of the nervous system.

In the following section, then, we shall be looking at electronic techniques as the first stage of automation of primary economic function in the living entity we know as 'Earth'. We shall explore to what extent computers and telecommunications can be used to develop a para-nervous system for the planet, thereby furthering its development as an organism and enriching the lives of men.

I foresee that in the course of time such a system will be superseded by intuitive organic operation involving direct brain to brain contact between men, but we're not there yet. Nor may we wait until men become more intuitive before we put our house in order. We have to use the tools immediately to hand - tools whose integrating potential we've barely begun to explore.

In the next chapter we shall be exploring relevant capacities of the new technologies, and ways in which they may be able to assist us in forming viable economic relationships.

CHAPTER THIRTY ONE

7.5 Toward a Para-nervous System for Planet Earth

Summary: Exploring aspects of computer technology relevant to economic integration.

Advanced computer research today focuses on what is known as 'Fifth Generation' computers. It aims to produce a machine that adapts to the needs of its human user rather than demanding, as present-day computers do, that we should tailor our thoughts and instructions to suit the machine. Jack Weber explains that it will, like any good assistant, be there not only to perform the mental drudgery, but also to offer advice based on detailed knowledge of the subject.

> It will speak our language, whether that be English, French or Japanese. When faced with a problem, it will not ask us how to tackle it, but will work out its own strategy. If requested it will, at any stage, explain its reasoning in reaching a particular conclusion or adopting a particular line of thought.[1]

One of the characteristics of the Fifth Generation is that it will rely very heavily on knowledge-based reasoning, a capacity clearly essential when it comes to integrating economic requirements with the sustainable use of the environment. We shall be exploring this aspect of computer use in some detail later. Sufficient to note here that in this respect it comes at exactly the right time, when the knowledge available to humankind is so extensive it is quite beyond our powers to integrate on conventional lines.

Of the two main directions on which Fifth Generation computer research is focussed at the present time (perceptual vision and language), it is computer vision that shares this difficulty with us in the kind of knowledge-based information exchange I have in mind.

> Essentially the problem that looms over all attempts at computer vision is one of sheer volume of information. The richness of detail that we can see, and need to see, makes even the best computer systems seem pathetic by comparison.[2]

However, there seems to be little doubt that this problem will be overcome, at which point the capacity of Fifth Generation computers should be entirely capable of integrating our material activity on a global scale. Activity will be generated locally, and fed into the system in a comprehensive computer-linked network for wider processing and feedback.

In this way it will be at all times the quality of individual input which determines the global flow. The system I envisage here is comparable to the approach of 'Fifth Generation'computer research known as 'bottom-up'design. Except in a very general way, we cannot explain how it works. As in the human brain, its abilities come from inside itself and do not lend themselves to sequential analysis.

Jack Weber explains that top-downers start from the overall task and attempt to write a program that will perform it.

> What they are producing is programmed intelligence - algorithmic or heuristic techniques which are designed to achieve some result that is also produced by the brain.[3]

Weber notes that the program may, of course, be quite different from the pattern of mental events, but it can nevertheless provide valuable information about what is necessary and what is impossible in a particular process. For example, if a program that is designed to interpret a three-dimensional object from a two-dimensional drawing gives ambiguous results it might be quickly ruled out as inadequate. But if its failure corresponds to a known optical illusion, it may be a useful experimental tool for investigating the reasons why we sometimes get confused.

Bottom-uppers approach the problems quite differently, building on the knowledge that the brain is organized as a vast ever-changing enmeshment of neurons, each of which is interconnected to as many as several thousand others. Weber observes that whatever is going on in this network is obviously a very highly parallel process, quite different from the workings of any normal computer.

> To try to mimic such activity with a sequential program would be a hopeless task; instead they ignore the details of what is happening and aim to imitate the complexity of the structure. Its behaviour may be difficult to predict in detail but it can be made to respond very intelligently.[4]

Researchers know that the two techniques are not mutually exclusive, and that both probably have counterparts in the working of the brain. They therefore believe that any sophisticated computer system will eventually have to do the same.

In attempting to apply these techniques to society, we find that in the human situation, hidden variables are virtually infinite, so that it may be the bottom-up approach that is likely to serve us best. The overall Gaian impulse will direct the result. That this impulse will be for good, rather than ill, is a trust without which we might as well pack up before we even begin. It rests on the spiritual nature of the universe, and stems from God.

At a practical level, self-regulation of society with the aid of computers must proceed from certain basic assumptions:

- that most individuals would rather eat than starve

- that by virtue of being human most individuals are capable of some contribution to society, however low they may have sunk, no matter what their physical handicaps, their moral degradation or their past record, and that most, given the opportunity, would prefer dignity to degradation

- that most individuals are capable, under the right conditions, of loving something or someone

- that most individuals are disillusioned about the merits of existing systems

- that most are capable of response to hope concerning their own immediate fortunes

- that, properly managed and integrated, there is sufficient potential on earth, combined with ever-increasing skills and knowledge, to provide every contributing individual with basic material necessities, provided that we keep population growth within ecologically sound limits

- that, properly integrated, the provision of these basics is obtainable by the skills and will of the people who need them

- that there are a sufficient number of loving, motivated people on Earth to carry human vision beyond the necessity for bread.

What, then, do we mean by 'properly integrated'? I think it means this:

- that the system as a whole should be open, at any and every moment, to the need and potential of every person in it

- that it should be programmed to be responsive to this need and to engage this potential

- that it should be entirely fluid in operation, allowing for continuous adjustment and change, but holding society in dynamic, happy equilibrium

- that it should be programmed within the context of a living Earth.

If the flow of goods and services is to proceed from the self-motivated activity of millions of individual men and women, in such a way that, with the aid of computers, it is self-regulated and sustainable, are there any identifiable needs common to all human beings that can guide us in moving towards such a society?

We have identified a number of principles that seem to emerge from a study of multicellular living systems. We have also looked at certain imperatives emerging from these principles that seem to have relevance for human society, thus characterizing the desired nature of that society. But if the new order is to emerge in response to each individual's loves and needs, and through his active participation in society - if, in other words, society is to be the self-regulated resultant of spontaneous interaction between people, at what point must their separate value judgements be brought into the system?

The answer is, I think, linked to another:

What do people really *want* of life (apart from bread)? What are *the* basic criteria of human psychological satisfaction?

I think we need three things: firstly, a sense of personal identity, personal worth. Secondly, a sense of belonging. And thirdly, a transcendental goal. Because these needs are common to all mankind, there will be a sufficiency of cohesive force to hold the whole thing together.

CHAPTER THIRTY TWO

7.6 Towards a Non-monetary Free Market System

Summary: The new information technologies can provide a viable substitute for money, in a system still to be regulated by supply and demand.

Here are some specifically economic considerations that must govern the programming of computers in their integrative social role.

- The input of individuals, groups or companies should be assessed for their impact on Earth's capacity to renew herself. Thus high-pollutionenterprises would be unacceptable, as would be excessive exploitation or needless destruction of life and resources.

- Practices in which the prosperity of one is obtained at the expense of others would be unacceptable and would be phased out, in recognition of the fact that since the Earth is finite, over exploitation by one group may mean not only harm to another but the surcease of all.

- Growth at all costs will be seen to be self-defeating and inappropriate, since a living economy encourages only such growth as is appropriate to the needs and scale of the organism.

- The pursuit of satisfying activity will channel people, companies and groups into patterns of life as bewilderingly different as are the life forms and systems found in nature. Computers must therefore be programmed to allow for an almost infinite number of modes.

- There will be nothing static about this patterning. All will interrelate in an everchanging kaleidoscope of interdependent, reciprocal association in which, since all are part of the same living system, each individual, each group, each industry and each enterprise will ultimately affect all, either directly or indirectly.

- Each group, each production unit, each household or farm or clinic or clearing house will find that it has an optimum size like the whale or the worm or the robin, because of the mathematical relationship between size and function.

- There will be no waste. Each functional group or subsystem of whatever kind, but especially in industry, will be 'fed' from the by-products of another, and its own products or its own wastes will in turn be used in a related life process, as is the case in natural systems. Computers will be programmed to correlate such activities.

- The computer will make available to any individual, any group or any enterprise the latest and most relevant research in his field, at the same time feeding back new field-research data into the memory store for further analysis and integration.

- Individual initiative and participation will be paramount. The goal will be full integration of each man's full potential.

- Each group, each industry and each enterprise will preserve a kind of homeostasis by a continuous process of adjustment appropriate to the scale and nature of the enterprise, ensuring thereby dynamic equilibrium between the 'organism' and its environment.

- The only requirement for access to the system will be useful activity in the context of a living Earth. This will ensure a free flow of any required goods and services that are available. No money is required in such an exchange.

The idea of a moneyless economy is of course not new. What is new is that it should be within the competence of the individual to free himself from dependence on economic manipulation, whether of 'free market'or of bureaucratic origin.

Under communism, it was envisaged that goods and services were to be distributed according to need, rather than in relation to work performed, and it was argued that the worker would not require monetary payment, while work itself was to be performed more as a social duty than for purposes of individual gain.

One would need to more than starry-eyed to believe that a system like this could work without coercion. To begin with, who but oneself can assess one's need? and by whom is the allocation of goods and services to be decided? If by some outside body, a degree of dissatisfaction is inevitable. Hence coercion.

In the matter of work, the percentage of people in any society prepared to work as a social duty is always grossly outnumbered by those who need the motivation of individual gain. Laudable though the basic idea may have been, it is, I think, incompatible with organic principles, in which both need and output are self-initiated according to the character, physiology and function of the individual, whether it be a single cell or a more complex organism.

Attempts to run an imposed moneyless economy have run into some strange anomalies. Marx suggested that under socialism workers would receive 'vouchers' entitling them to consumer goods. These vouchers would be proportional to labour time. Since these vouchers did not circulate, he argued that they could not be considered as money.

Another strange aspect of the 'moneyless'economy concept was that when it was finally abandoned in the Soviet Union in favour of a system that retained some private enterprise accounting, and the payment in some instances of money wages, there emerged a concept of so-called 'passive money', described by Alan Smith as follows:

> A simple definition of passive money is that the possession of money alone does not enable the owner to acquire resources or goods and services whose allocation has been determined by a plan instruction. Passive money therefore functions purely as a unit of account, and money flows follow, rather than determine, resource flows.[1]

When juxtaposed to the function of money in a capitalist economy, this notion presents a startlingly different role, since in a capitalist economy, capital-and-money-flows clearly dominate resource-flows. Money in the West spells power to produce. 'Passive' money in a centrally-planned economy was intended to supplement State control.

But do we really need this means of exchange, the way things are in the 1990s? Is even 'passive money'necessary or helpful? At the time when Karl Marx and his followers were wrestling with these ideas, electronics had yet to be developed. The possibility of a direct transfer of goods and services without a tangible means of exchange was still far in the future. After a while the planned economies found it impractical to do away with money altogether, and modifications made in 1939 provided a clear distinction between cash paid for wages, and cashless transfers between enterprises. The latter were obliged to follow plan instructions from the State, and transfers of material inputs were 'debited and credited accordingly by the State Bank without any actual cash transfer taking place.'[2]

This involved the creation of two distinct and separate monetary circuits running side by side, an encouraging pointer to possibilities in a transition period. It seems likely that in transition to the information-based economy I envisage for the future, the concept will be extended to allow not just for two, but for many diverse and complementary conditions of exchange. It is clear that the adjustments facing Germany since the unification of East and West, and those presently facing the Soviet Republics since the collapse of communism in 1991 will prove intensely interesting proving-grounds for new ideas.

However that may be, in today's situation it is a gross oversimplification to think of the world in terms of communism or free enterprise. Free enterprise will find a way to intrude in the most determinedly socialist society, and there is a great deal of communal activity in matters of common concern in the West, in spite of the current tendency to privatize previously State-owned enterprises.

With established patterns wearing a bit thin, people are beginning to want not so much an ideology as *a good life*. Increasingly they find no nation-state, no political party, and no ideology able to deliver it. But this is not the end of the story. The potential for integration by the new technologies has yet to be understood. When it is, things will move more swiftly than we can dream.

CHAPTER THIRTY THREE

7.7 New Trends in Economic Thinking

Summary: Capacities and limitations of computers in giving expression to the new trends in Economic Thinking.

A number of new trends have emerged in recent years in economic thinking. Those relevant to a living Earth include:

- that people matter
- the necessity for a sustainable Earth
- the value of diversity
- the cumulative value of informal business
- the job-creating potential of small business concerns
- the need to draw into the system large numbers of people presently living on the edges of society
- the value of previously unpaid services, including home-making, the rearing of children, and care of the sick or aged.

Here are some of the ways in which the new technologies can help us to implement these important new ideas: They can:

- enable us to think globally while acting locally
- accustom us to thinking appropriately through more comprehensive access to the relevant facts
- integrate local activity to maximum benefit for all
- free vast sections of the population from gruelling or repetitive labour, and channel them into self-chosen integrative activity
- integrate the illiterate millions into the system through direct educational facilities
- make the small man aware of openings for his skills
- increase productivity through integrating personal motivation
- create numerous new service and information-related industries
- facilitate recycling - eliminate waste through access to knowledge concerning the use in one industry of the by-products of another
- reduce costs by making advertising unnecessary
- flash new insights and information around the world in a moment of time, so they become the immediate property of everyone and are

integrated into the world store

- link everyone with anyone, at will, so that trade links are easily formed

- make everyone aware that the elementary criterion for survival, economic or otherwise, is that their thinking be global in character and intent, and that it be directed toward a sustainable Earth

A number of interesting developments will spring from these possibilities, and while not strictly economic in character, they will have immense value in the general contentment of the population, and will therefore reflect in the economic performance of its members.

By employing a large part of the population in humanizing cerebral activity, we shall, by using our intellectual and emotional capability to the full, heighten the threshold of human thought, setting in motion thereby a renaissance of learning such as the world has never known. Everyone's creative thought will be enhanced through his accessibility to the world's store of beauty and knowledge, and its application in his individual sphere of interest.

Moreover, by making demands upon us for the production of ever more software, and keeping our minds thereby at continual stretch, the new technologies will ensure that new functions of the brain will be brought into play and new capacities developed. They will also ensure that the individual dynamism of one man can inject an immediate resonance into the matrix.

Here are some of the things computers *can't* do:

Feel, marvel, worship, care, grieve, rejoice, hate and love.

They *can* think, relate and transcend, which is only natural, since they are creations and extensions of our own minds, but they remain tools, as fully as a hammer in the hand or wheel on the road. Much of the controversy surrounding them springs from a misunderstanding of this simple fact. In all our dealings with computers, we need to remember this. Computers will do what we tell them to do, and what they can do is immensely useful to us. We need to use them to the fullest stretch of their capacity, extending at the same time those imaginative qualities of mind and spirit in which they have no part.

If we are to help people, and if we are to be a positive force in the coherence of the Gaian entity, we must not lose sight for a single moment of the fact that every cell carries within it, in latent form, the whole of the organism's potential responses. It is therefore perfectly possible for us to seek and find within ourselves a vision of her emerging selfhood, and of the ways in which we must think and act to further her on her way.

Moreover every cell is, in multicellular organisms, in direct relation to every other, by means of the blood, lymph or nervous system, or by means of hormone secretion in the ductless glands. This is why, in nature, the astonishing things happen that do happen. This is why multicellular organisms, of whatever complexity, really work. And if we find this process difficult to envisage, we should remember that the dynamic of intercellular relationships mirrors exactly the principles of subatomic physics discussed right at the beginning of this book: Every particle is in direct and continuous relation to every other anywhere in the universe. Similarly every cell is in direct relationship with any or every other in the organism. By developing the technologies already available, every man may now be in direct relationship, at will, with any or every other anywhere on Earth, and will, in any case, react upon them, whether he will it or not. For the very first time, then, something akin to synthesis is within our conscious reach.

Here I must insist again that I do not envisage electronic synthesis as the be-all and end-all of the human communication facility, nor even as a permanent tool in the Gaian cell. Electronic devices may be, at best, just our clumsy beginnings. Already there is talk of the microchip giving way to the 'biochip', in which proteins are used as the material of computing, and the fast growing science of genetic engineering may make possible, in the not too distant future, yet more fantastic advances in computing capability.

> One suggestion is that cell membranes - the flexible walls which surround every living cell - could be used as an effective computing material because they already contain very complex chemical structures for controlling the distribution of electrical charges. There is even research into methods of connecting living nerve fibres directly onto semi-conductor chips.[1]

Beyond these again is the cerebral reach of man himself, which must finally transcend his own creations. Whether he will then supersede his marvellous tools, ordering his corporate life in subtler ways, remains to be seen. But we cannot wait for these things. We have to go to work urgently. *We have tools enough for a start.* It is in that spirit that I would wish to explore their usefulness.

Book II Part Eight

Microchip Economics in Action

CHAPTER THIRTY FOUR

8.1 Economic Applications of the New Technologies

Summary: The new technologies are found to facilitate economic performance at local, regional and global levels.

'Ideas are reality of a very tough texture indeed.'

Stephen Bodington, *'Computers and Socialism'*, (Bertrand Russell Peace Foundation. Nottingham, 1973)

It is time to take a closer look at some of the ways in which the new technologies can be applied in the economic field.

At a local level they can:

- Inform us, i.e. give us knowledge immediate to our needs and extend our grasp of available resources and potential.

- Keep us in touch with opportunity, i.e. advise us concerning the needs of others in our area in:

a) Goods (leading to production of some kind).

b) Services (leading to useful activity of some kind).

- Keep others in touch with our needs or skills in these areas.

- Keep up to the minute tags on *what* is being produced, *where* it is being produced, *by whom* it is being produced, *how* it is being produced, and what the producers require in materials or labour or know-how to further the usefulness of their production; what is the nearest source of such materials or labour, and whether further centres or factories are required in that area to produce more of the same product. Also what other production could usefully be set up to complement and enhance those already in existence.

- Keep up to the minute tags on what people need or want that is *not* being produced, what areas would be most suitable for production of these items from the point of view of resources, skills, labour, ecological stability and distribution to complementary industries.

- Keep up to the minute tags on what is being *grown*, and where; how crops are progressing; whether they are adequate to people's need; whether they are the best crops for local conditions of ecology and climate; whether they complement other agricultural activity in the area; what other crops could be grown with them to enhance ecological stability, nutritional value, or reciprocal exchange with adjoining areas.

- Monitor and advise us on pollution levels in our area; on their source, on the best methods of improving them, and on the appropriate contact to make if the source of pollution were found to be further afield.

- Advise us concerning useful applications for the recycling or disposal of our waste products, as for instance in the recovery of metal scrap, or the biological production of energy.

- Correlate research on local ecosystems and their usage and preservation, and make us aware of economic opportunity in this field.

- Keep us informed concerning local facilities in recreation, health care, clubs, hobbies, educational facilities, housing etc. and of the economic and service opportunities in this field.

- Really find out, for the first time, and monitor continuously, what each person:

a) is doing
b) would like to do
c) is able to do
d) has need for . . .

. . . and correlate these needs and affinities.

At a wider regional level they can integrate all the activity and information just cited, and extend it to include county, national or global activity in any desired field, as well as facilitating the most suitable and economical transport over long distances, and between diversified areas.

In the global ecological field *on which the economy finally rests*, the new technologies can:

- Monitor the ecological health of life-support systems both on land and sea, keep us informed, and correlate corrective measures.
- Monitor resource levels and keep us informed. Correlate usage levels at sustainable levels.
- Monitor population levels and densities and relate them to carrying capacity.
- Monitor energy requirements and availability, and correlate research into new and related sources of energy. Pinpoint suitable sources of energy for local areas and advise on them.
- Monitor geo-physical phenomena and their applicability in related fields.
- Warn in advance of natural disasters, where they will occur and when, what their extent is likely to be, and what emergency help may be required beyond the local resources of the area concerned. Correlate manpower and resources to meet the emergency.
- Keep constant tags on weather patterns.

In the field of labour and manpower the new technologies can:

- Bypass illiteracy, so that visual and auditory channels can keep people informed.
- Bring isolated rural communities into the body corporate with all its benefits.
- Assimilate into the body corporate the products and wisdom of these isolated communities.
- Bring education within the reach of millions of illiterate people.
- Bring health care, preventive medicine and know-how to isolated, illiterate or rural communities.
- Afford such people access to the training and skills which will enable them to help themselves.
- Absorb millions of now-unemployed intelligent people in training in computer skills.
- Where people are unsuited to this kind of training, or have no liking for it, give them access to the training of their choice in other fields.

That the new technologies are here to stay is very clear. They have invaded commerce and industry, and wield increasing influence in almost every aspect of social life. There is resistance by people fearful of change, and outcry from others who fear a dehumanized world. Neither succeeds in making the slightest impression on computer growth. You cannot un-know what you know. As well try to stop the usage of print, or the motor car. It just can't be done. And we do need help. Computers can offer exactly the help that we need. They are not alive, but neither are they only machines. They are instruments to integrate our separate minds; tools to extend them. Rightly used, they can help us to humanize a grossly inhuman world.

This is a very far cry from the way we've used them so far. The firms introducing automation and computer techniques most rapidly were predictably the richest and commercially strongest, and used the new techniques to gain yet more control. In addition, they could obtain the services of the most highly trained people in the field, keeping in this way well ahead of lesser concerns.

From the beginning, then, the new technologies have had exactly the reverse effect to that envisaged for a living Earth. Instead of diffusing money and power towards a more general good, they have reinforced the tendency to concentrate it in fewer and fewer hands, bringing the world economy to the verge of breakdown in the present decade.

However, this is not the end of the story. Patterns change. Today's computers are smaller, cheaper, more efficient, and more generally accessible. Far more people understand them. Children are growing up with an apparently in-built facility to interact with them, even at pre-school age. Computer games, computer toys, computer graphics and a host of other applications are part of the daily life of millions of youngsters, rapidly producing a species of man to whom they are simply a part of living.

It has been noted by many that man creates technology, which then recreates the man. Conditioned from infancy to a high-technology environment, it will be no more than natural to the next generation to use these tools, and to regard computerized information-processing as the natural regulative activity of the planet.

As I envisage it, this activity will engage a web of information channels so well articulated, and so widely diffused, as to integrate all aspects of goods-and-service exchange in a dynamic flow. It will need neither money, nor vouchers, nor tokens of any kind. The prime regulating factor, subject only to imperatives of global well-being, will be the information fed in by each individual or group concerning his own needs, capacities, preferences and inclination, all continuously and changingly matched with an up to date assessment of local potential, and thereafter linked in an ever increasing web of human 'ecosystems' right to the global level.

Information will be fed in by each, available to all, and controlled by none. All networks will be further enmeshed with data banks of all kinds, using telephones, cables, radios, television and satellite links, and any other technical means available, thus bringing the world's full store of knowledge, research and development within the reach of all.

Enormous numbers of dedicated people will be required for the programming of these computers, and for the collection and correlation of information in data banks. We shall need to transfer to computers the wealth of knowledge on our library shelves. We shall need to link libraries, universities, research organizations, observatories, hospitals, schools and private homes in a network so close that everyone is part of a dynamic, active interchange of thought, continuously updated. We shall need to put people intimately in touch with other people in any field of their choice, whether they be biologists, ecologists, doctors, teachers, psychologists, social workers, manufacturers, ship builders, lumbermen, farmers, artists, housewives or musicians, or any other of a thousand unnamed groupings that make up a society.

Programming should be biased toward a sustainable global economy, preservation of the habitat, species diversity, renewable resources, pollution-free enterprises, recycling, reciprocal industries, individual initiative, full use of human potential, flowering of personality, easy access to information, self-help, mutual support, healthy living, artistic expression, spiritual renewal, challenge, respect, and a turning to God.

> We are talking about the possibility of everyone's energy, ideas, knowledge and skills being used creatively to provide what our human society needs and wants.

The above are key words in Stephen Bodington's book, *Computers and Socialism* published by The Bertrand Russell Peace Foundation Limited in 1973. Although I only came across Bodington's work quite recently, and though I hold no brief for socialism, I have found a great deal in it that complements my own. I acknowledge and thank him for it. His ideas will be discussed in some detail in the next chapter.

CHAPTER THIRTY FIVE

8.2 Down to the Nitty Gritty

Summary: There is no technical reason why information-exchange should not replace market-exchange as a basis for economic activity. Difficulties are likely to be social in nature rather than technical.

Bodington's book, *Computers and Socialism,*[1] is a masterly analysis of the way things actually work in the bewildering world of economics, and as such has been helpful to me. Bearing in mind that the book was published in 1973, I find his insight remarkable, for the possibilities inherent in computers and telecommunication have increased exponentially since then.

Bodington observes that there is no real technical reason why information exchange should not replace market exchange as a basis for economic activity, and that the difficulties may be social rather than technical. This is correct, I think, for even given that there is enough to go round, equitable distribution comes up against our sense of social justice: Some people and some groups will have more to contribute than others.

Further to complicate matters, and starting from where we are now, some people and some groups will have a flying start, a source of wealth and power they will not happily relinquish. Both these factors are formidable barriers to equity. But if under a different system, we are able to have, and to be, whatever we wish, and if there's enough to go round, do we still have need for our present concept of wealth? *Must* values still be assessed in terms of price? Do we actually still *need* money at all?

In accepted economic terms, we do. *If Earth is a living creature, we do not.*

As already discussed above, equality of value as a prerequisite for benefit is simply unknown except by mankind. What we now need instead is the efficient coordination of dedicated human minds. We can achieve it by means of computerized information exchange, articulated in such a way as to ensure a continuous and adequate flow of basic needs. And this is nowhere more urgent than in the field of economics.

In any economic organization there is a need for a daily and adequate flow of information: between suppliers and producers; between producers and those who market their products; between marketers and users.

Producers need to assess the needs of the market, their capacity to respond, and the availability of raw material and equipment. Bodington calls this 'a sequence of information flows' which will 'communicate awareness'.

An interesting feature of this author's thesis is his insistence that the function of each organization needs to be tested for its ethical/social/aesthetic value, and that economic values alone are not adequate.

> Each organism requires greater power to shape the pattern of its own achievements and at the same time greater discrimination as to its function in the wider pattern of socio-economic organization.[2]

He puts his ideas in a nutshell with the following words:

> Linkages which hitherto have been made through the market may now perfectly well be made through computer-based information systems. This is the crux of the matter.[3]

Bodington's reference here to the term 'market' equates it with the monetary system, which is probably in line with the generally accepted meaning of the term. My own usage, on the other hand, extends the concept of market demand to include what people need or want, not just what they can pay for. However, even in the narrower conventional sense, there are a number of ways in which computerized linkages are superior to market-related information.

To begin with they can more readily be extended to individuals, groups and production units having no capital save their brains and hands, thus ensuring appropriate activity from grass-roots level. And since no individual and no group can complete all aspects of production of even the simplest industrial article without taking something from someone else, a maximum exchange of ideas, skills and capacities between all the people involved is essential to maximum efficiency.

Computer linkages of the kind envisaged have also made possible the satisfaction of a man's basic human need for self-chosen creative activity, since his own circumstance and that of all those with whom he has social or economic relations become known, each to the other, at will.

Then there is the question of speed. No-one knows better than the industrialist the time-lag involved in relying on market performance to inform him whether his product will sell, nor the risk involved in inaccurate forecasts. Electronic processing of information now makes it possible for anyone to obtain immediate information about the requirement of consumers, as well as about sources of supply, availability of labour, the most up-to-date methodology and so forth, and to make known his own response to such information. No matter how complex the data involved, provided the computer is properly programmed, no time-lag is manifest.

> The life-blood of effective relation is full, suitable and rapid information. Any work team or consumer group in doubt about what it should do can only resolve its doubts by asking questions and getting answers.[4]

Bodington points out that it is just in this area of rapid understanding and interpretation of changing situations that the computer revolution makes new facilities possible.

> To exploit these facilities implies both a clear vision of what is wanted socially, and energetic development of the potential that is already there.[5]

The new techniques of information-processing have been employed so extensively since these words were written, that industry would virtually grind to a halt without them, while the social vision which could make them ethically acceptable has lagged behind. We need to remedy this, and we can. The character of the new tools will facilitate immediate communication, as well as the establishment of non-competitive, mutually acceptable arrangements in commerce, trade and industry. A more satisfactory social ethic must surely follow.

Rural applications

In rural situations there do not seem to be insurmountable difficulties in the way of ordinary people acquiring the basic necessities of life, using information as their means of exchange instead of money. Reciprocal help works well enough, especially in small rural communities and villages.

If, for instance, I need a house, there will often be people in my own immediate group ready and willing to help me build, once the pressure is removed from them of trying to procure enough money for their own needs. Should there perchance be no one in my immediate community able and willing to assist me in the building of my house, or in the making of bricks, or should there be no timber, or piping or glass immediately available, recourse to my local computer terminal could both make known my needs and inform me of the nearest point of access to such materials or labour, as well as advising me what needs, goods, labour, knowledge or skills would be most likely to be reciprocally useful to the people able to supply them. Should I not be in a position to contribute towards such needs, someone else in my group might be. And I might contribute usefully to him in some other manner, or to his friend, his child, or the group as a whole. Oblique barter of goods and services is in this way enormously more flexible than direct purchase in money, since nine tenths of society does not have money in sufficient quantity to make our market system viable. *To allow one single factor - the monetary means of exchange - so to bedevil the whole organization of society* is manifestly absurd, especially now that we can replace it more efficiently.

To make an information-based economy work, then, will be more a question of getting used to new ideas than anything else. In addition it will require that great numbers of people be trained in computer usage and programming - including those in rural areas; that access terminals be widely linked and widely diffused; and that increasingly specialized data banks be set up in each area to record the new experience and specialized wisdom of that particular community, and to feed it back into the network.

Industrial application

When it comes to industry, the situation is much more complicated than in rural pursuits. Industry and the market are blood brothers, so to speak, and not easily separated, even were it desirable to do so, which it is not. What is envisaged is an adaptation of market structures to accommodate the flow of goods and services in a manner not dictated by money flow. Should this moneyless exchange by means of adequate information prove to be practical and satisfying, the existing industrial structures could usefully benefit the many instead of the few, through the unconscious self-regulation of production by a network of interlinked computers and data banks - such banks to be freely and randomly activated by a pervasive flow of information between all members of society.

It is important to note that this does *not* mean the regulation of production and distribution by an industrial elite empowered to control the computers, nor by any kind of government agency. In the organic context it is quite specifically *self-regulation* at all levels, and is at no point centrally controlled. In Bodington's words:

> The potential of modern science and technology can only be fully realized in a system in which decision-making and information are widely diffused.[6]

If information is to perform for society the function that money now performs in the market, we need to look very carefully at the linkages that actually occur in human activity - in agriculture, in commerce, and in particular in industry, so that those people who affect each other's lives most directly in the course of their work may pool their knowledge and skills and program their local computers effectively.

It can hardly be over-stressed that the greater the complexity of, say, an industrial undertaking, the more comprehensive must be the exchange of information if the whole structure is not to fall apart. We have already seen that modern computers may soon be able to take in whole situations 'at a glance' in something of the same way as that in which an eye sees, and may moreover be able to supply selected output data conforming with any specified pattern or code. The kind of complexity evidenced in modern industry is therefore unlikely to present insuperable difficulty from the point of view of data processing.

Coming to grips with the ramifications of human activity in the industrial process, I cannot do better than once again to let Bodington speak for me. The example he gives goes back to 1948, but is none the less illuminating for that:

> As a starting point, and in order to suggest some aspects of the reality from which we are abstracting, consider some of the economic relations implied in a common product, such as might be purchased at any time during the last thirty or forty years in an industrially advanced country such as Britain. An electric light bulb will serve as an example.[7]

Bodington lists the main components of the common globe: a bulb, a cap, an exhaust tube, a glass support, a rod, leading wires, filament supports and a filament. The cap is made of brass, an alloy basically of copper and zinc. Copper wire is used for the lead-in wires, molybdenum wire is used to make the filament supports, tungsten wire to make the filament, glass rod to make the support which carries the filament supports. The product thus requires materials from all over the world, and the combination of a variety of manufacturing processes. To be economic, production must be on a very large scale involving heavy installations, furnace equipment, bulb blowing machines, and other factors.

The manufacturing process is preceded by the manufacture of tungsten and molybdenum wire, and the precision operation of filament coiling, as well as by other intermediate processes. The whole process adds up to a situation in which the individual worker is unlikely to know, let alone control, all the processes involved in producing the electric bulb.

> But what he does, and the payment he receives, as a worker, establishes in fact a very concrete nexus of socio-economic relations with other people throughout the world. [8]

At the time of which Bodington is speaking, each worker filled a specific position in the production line, such as minding machines, packing products, writing invoices, managing the sales force, determining production schedules, and so forth. Fortunately these soul-destroying methods of mass production are giving way today to a truer 'team' concept, in which workers interchange function much more readily, and in which they are encouraged to become familiar with the whole process. But either way, the production team has specific and wide-ranging relations with those who produce their raw materials (molybdenum, tungsten, etc.), with the manufacturers of the machines that they use, with the producers of all related components, and, in many instances, with those who market their product.

Bodington notes that market links are indissolubly intertwined with relations to other groups of people - through the money system that mediates sales and is the source of money capital. Also that capital itself hides a host of relationships between people as owners and employees; as those who pay and those who receive interest; and with other forms of income distribution. Capital structure also determines the distribution of resources, what industries are to be developed, the level of wages and salaries, and the flow of trade.

It is clear that these monetary transactions form the basis for a multitude of relationships between individuals, suppliers, manufacturers, packagers, distributors and consumers, and that in so doing they determine to a large extent the patterns of behaviour through which they are implemented. *Bearing in mind the highly competitive nature of the business world, it is hard to imagine a more unfortunate basis for the development of human ethics*.

We are aiming, then, at a situation which might help to remedy this unfortunate bias in market relationships - a situation in which, not only will the speed and efficiency of these exchanges be enhanced, but the relationships between people may become more direct. In this way the organization of society will once again embody the face-to-face nature of earlier and smaller communities - and this despite the restraints of distance or the overwhelming numbers of people involved.

We are moving toward a situation in which, with the aid of the new techniques, we may once again be able to see and talk to the people with whom we are involved, and in which the processes of production and marketing become, first and foremost, a matter of interpersonal caring relationships.

We all know already how closely we can relate to a screen or television personality, to the point that we become unaware for the time being that we are looking at a pictorial image rather than responding to a living presence. We can already attend conferences by means of the television screen, extending the face-to-face visual experience and enhancing personal involvement thereby.

As these facilities become more widely available, informal groups, and individual men and women, will begin to use them freely, so that anyone can 'see' the person or persons with whom he wishes to do business, or with whom he might like to work, and can talk about it informally person-to-person, without perhaps having to travel six thousand miles to do it. He will be drawn to one, repulsed by another, trust or mistrust a third, feel doubt or confidence or even friendship for another, as readily as he would were they together in the village of the tribe.

Given new attitudes of people and the phasing out of money as the sole arbiter of economic activities, organizational problems will thus resolve themselves by the establishment of multiple information flows between consumers and producers, most of whom can become personally known to each other at will. The resulting stimulation of initiative, motivation and skills must be beyond price. The quality of society can only be enhanced.

Where then do we start? Clearly from where we are now, in what Bodington calls 'the actual historical present.'[9]

For the past fifty years the world has operated on two major and conflicting economic ideologies, roughly corresponding with East and West, with a number of hybrid viewpoints between. In the East, various forms of communism and socialism. In the West, the so-called free-market system, based on production for profit. The collapse of communism in the Soviet Union and elsewhere in August of 1991 has given a tremendous boost to proponents of the free-market-system, but we have to ask ourselves seriously whether, in its present form, it can feed the world's starving millions or put the ecology back on track.

Both systems are profoundly anti-systemic; both are foreign to a natural-systems approach. But where the Western system scores is in its insistence on individual freedom and choice. This is because honest self-interest, which both societies deplore, is, in nature, synonymous with self-preservation, where it emerges as a very healthy biological principle. In nature everything earns itself the right to survival simply by indulging in those activities characteristic of its kind. Within the framework of a global ethic, the principle holds for man.

Self-interest has been condemned by man throughout history, yet it is practiced everywhere (usually under a holier-sounding name). And ironically, it is this very principle of self-interest in a transcendental form that may prove our salvation in the end. Only when it is pursued in opposition to the greater good of the whole - when self is opposed, as it were, to greater self - is it dangerous and lethal to mankind. Properly synchronized (as with the aid of computers and other electronic devices it now can be), and in full realization that individual and corporate well-being are now synonymous, it will be a generative force second to none.

CHAPTER THIRTY SIX

8.3 Day-to-day Situations in a Non-Monetary Economy

Summary: Computers are well able to correlate the daily lives of human beings to the benefit of all, without value assessment or the need for a monetary means of exchange.

I want to look more closely now at the ways in which the day-to-day lives of ordinary people will be affected by a non-monetary information-based economic system.

The most startling and beneficial effect will be the removal of anxiety and monetary rivalry in society at large, with untold psychological, physical and spiritual benefit to all mankind.

Next might come a reduction in wars and crime, in drug-related degradation, in violence, in vandalism, in child abuse, in environmental destruction, and in a thousand unnamed but interrelated ills to which society is prone.

However it is the economic aspect of all these things that is the subject of this enquiry. So we shall be concentrating on the ways in which the activities of the man in the street will be integrated into the 'economy', and on how he will go about attaining his daily needs.

We in the cities are so conditioned to a life dominated by the need for money, we forget that, in terms of population figures, we are outnumbered. So it is helpful to remember at this point Hazel Henderson's insistence that to equate society's total socio-economic productivity with that portion of it based on competitive, market-based cash transactions is a serious distortion of fact (see Chapter 25). She points out that most of the world's production, consumption and maintenance still occurs outside the monetized economies, and that our present system by-passes both the ecological basis of the economy and the countless unpaid activities of most of the world's population.

In spite of the terrifying disruption of rural life and the drift of millions of people to the cities, rural people still constitute a very large proportion of the world's population. No system that fails to integrate them successfully into the world economy can be expected to survive indefinitely, and conditions suggest an urgency in economic reform. We look first at possibilities in rural areas.

Information flow required in rural and nomadic communities will be very different from that of big-city dwellers, and computers will have to be programmed accordingly. Predominant needs in a rural community might thus be for education, environmental awareness, improved methods of agriculture and animal husbandry, the provision of reliable water supplies, medical care, adequate housing, clothing, and certain consumer goods from nearby industrial areas. Computers could readily monitor the requirements of these relatively self-reliant communities, and correlate wider community action (as, for instance in the provision of water to a dry area) where necessary.

The city dweller is a different kettle of fish altogether, highly specialized in his activities, and dependent for his very survival on the smooth synchronization of economic processes. If food supplies do not appear in the city stores for a few days, he goes hungry. If the shelves are empty for a few weeks, he starves. There are always, of course, the privileged few, who are able to keep their larders well-stocked, or, in emergency, get in a high-powered car and buy up supplies of food elsewhere. We are more concerned at this point with the ordinary citizen, who gets by quite nicely in normal times, but has no reserve or privilege in time of need; and with those whose daily existence depends on a loaf of bread. For them it is essential that goods are available in the right quantities at appropriate places and at the right time. In a monetary economy, this state of affairs becomes daily less easy to achieve, since the smooth running of supply and demand is distorted by people's inability to pay.

We shall be looking at a number of widely divergent societies and the ways in which the basic organic principle of oblique barter could be developed to accommodate their differing needs. The most familiar first: This is a community of moderately well-off city dwellers in a country whose economy is in a reasonably healthy state. It has the usual sprinkling of mink and manure elite, and an increasingly desperate marginal group covering the poor to very poor, as is found wherever people congregate in cities throughout the world. This flood-tide of urbanized poor requires in-depth discussion, so will be dealt with separately. First a brief look at the housewife or working woman of moderate to comfortable means:

Under the present market system the housewife visits her local supermarket for foodstuffs and other necessaries at intervals suited to her particular lifestyle. She selects certain articles, they are passed through a scanner or rung up on a till, and she either pays for them in cash, or has them entered to her account; or, if she offers a credit card, they are debited through a computerized banking system to her account in the appropriate bank.

Now let us look at the same process operating on the oblique-barter system, self-operated through the medium of computer processing.

Our customer selects from available supplies on the shelves just as she did before. She carries a 'credit' card, as now, entitling her to reasonable supplies of the items of her choice. Who will decide what is reasonable? We'll come back to that. When she presents her trolley at the till, a scanner will pick up the coded marking on the package as before, but this time what is recorded is a description of the goods, not a 'price', as well as any desirable information concerning source of origin, supplier and so forth. This information will be electronically conveyed, not to the customer's bank as a debit against her account, but to a centralized local computer, programmed to regulate the requirements and inflow of consumer goods to the district, and kept in touch with the stock of goods on the shelves of local stores by means of the process just described. This central computer is also programmed to order new supplies through computer links with the appropriate suppliers, to monitor any increase or decrease in the demand for, or supply of, specific goods, and to search further afield through wider computerized links in case of need. It is able to do this because, recorded in its memory bank, is all the relevant data concerning her needs *supplied by the customer herself.*

This data would include size of family, present occupation, usual consumption per week or month, food preferences, entertainment requirements and the like, all designed to enable the computer to ensure an appropriate flow of goods to the shop or shops most convenient to the customer.

This same data would enable the computer to 'ration' goods that were in short supply. Supposing, for example, that I were accustomed to consuming 500 grams of butter per week. There is a shortage. I panic, and lift from the shelves two dozen packets of butter instead of my usual one, so that I may store them against lean times. The computer, immediately matching my normal needs against the two dozen packets scanned, would be likely to flash a polite little note on the screen that might run something like this:

> Sorry, we have a problem. Butter in short supply. Please replace 23 packets so scanning may proceed. Thank you.

With others waiting behind me in the queue, I should have little option but to comply, particularly if some of them had been unable to obtain their normal supply as a result of my squirrel ancestry!

The same kind of system would be applicable with regard to the 'purchase' of manufactured articles such as refrigerators, stoves, clothing and cosmetics, and even motor cars. The available articles would either be on show locally in the appropriate shops, or information concerning where they could be seen would be on screen. People's response, wishes, comments on quality, suggestions as to improvements, preferences, and the urgency of their requirements could all be conveyed instantly to the appropriate factories and clearing houses, thus regulating production and distribution on the basis of factual rather than assessed need, and obviating the need for tedious house-to-house 'surveys' and the other marketing techniques in use today.

In a wider sphere of activity, personal data (self-supplied to the relevant local computer) might include such data as academic or technical qualifications; present occupation; present study; past experience (in anything from marine research to street cleaning); the desire for change to a different activity, desired sphere of work, research or study; a request for training; present involvement in voluntary social community service; artistic or musical talents; organizational ability; hobbies; state of health; immediate need for nursing or other care; housing difficulties; and so on through the whole range of any individual's personal needs, wishes, capacities and interests. This personal 'data bank' could be updated by the client from moment to moment as circumstances changed, in just such a manner as that of a monetary account, in the certainty that at no time would the client see flashed on the computer screen such words as now appear with despairing frequency:

Sorry! Insufficient funds available.

The computer would thus be in a position to correlate the needs and skills of the community at an individual level, and to supply information on the authority of the client to any person or persons whose available skills or needs were reciprocal to his own. Failure to dovetail suitable work/service/need requirements locally could trigger a computerized search further afield. It is difficult to imagine a situation in which so wide an integrative network could fail to find a useful and satisfying niche for everyone.

Computerized data banks of this nature could work equally well for firms, factories, agencies, associations, clubs or groups of any kind, the main difference being in the type of information supplied to the computer.

Thus a factory might feed in details of articles manufactured, production capability, stock in hand, number of persons employed, skills required of workers, materials required in manufacture, type of plant, current distributors or other relevant information peculiar to its particular business.

An amateur flying club might feed in such things as location of club, membership details, availability of machines, licence requirements, competition dates, or a request for ground staff or flying instructors.

An association for helping the blind, on the other hand, would be likely to supply data concerning the number and location of blind or partially sighted people, the degree of help required, requests for volunteers, information on training facilities and job opportunity, availability and care of guide dogs and the like.

With all these computerized data banks interlinked, cross-reference facilities would give everyone up-to-the-minute information in his particular sphere of interest or need, from buying a new machine to finding someone to mind the baby, and the net result would be a wholly spontaneous, coherent, unconsciously integrated society.

It could be argued that people will feather their own nests by feeding in false information. I am under no illusions about this. People undoubtedly will if they see profit in it. But this is a moneyless economy, which restricts the motivation to less tempting spheres. Moreover a system as diffused and closely enmeshed as this would cause false information to rebound on the individual or firm like a boomerang: false academic credentials would fail to correspond to university records; false data concerning experience would conflict with factory or firm employment data; false statements from, say, a builder, concerning assistance rendered to the homeless might meet with a discreet but embarrassing flash on his screen:

> You appear to have built ten houses in three days for people whose names do not appear in community listing. Please rectify data.

Or a man twice convicted of theft who requested employment as Keeper of the Crown Jewels, could feed into the computer the finest false credentials in the world without the slightest hope of an interview for employment in his desired sphere of work. His screen might flash back the message:

> Regret position unavailable. Excellent pickings at de Beers Diamond Mines.

Frolics aside, false information could only discredit its instigator in the eyes of society, while profiting him nothing. After a few false starts, I think he might opt for truth.

It might appear that the system would break down because of the unreasonable demands of unreasonable people who wanted much and contributed little. Supposing that I want, for instance, in addition to my simpler needs, a speedboat on an ecologically fragile river, a sea-going yacht for weekends, and a high-powered luxury sedan in an area where roads are mere tracks through virgin bush.

The first thing to remember is that there is no person and no law, to supply or withhold these things from me. Everything is a matter of arrangements between people, of the manufacture and distribution of legitimate needs of the populace, and of my own skills, capacities and loves.

It all starts in small local groups of people well known to each other, who will have programmed their computers by mutual consent to suit themselves and each other, and this programming will vary in different localities, under different conditions, and in different groups.

So the coordinating computer in the area where I live, having been programmed with a bias toward preservation of the habitat, might suggest that speed boats were inappropriate on that stretch of the river, since they would endanger the life of the rare blue crane. I might be told the whereabouts of the nearest speed-boat enthusiasts, who built and plied boats on a more suitable stretch of river, and who were in need of skilled hands. The choice would then be mine either to stay where I was and forget about speed boats, or, if boating were my central interest, to move to the new location. In the latter case, it would be up to me to prove my worth in the new community, render such service as I could (perhaps in quite a different field of need) and become part of the boating community.

I might be told that sea-going yachts were still on the luxury list and (if the economy were still partly monetized) were obtainable from such and such a firm at xxx million dollars; or, if we had already moved to a totally money-free economy, the same kind of choice would be open to me as with the speed boat. My capacity to have the things I desired would depend on their availability, on my own skills, and on my value in the community of my choice.

We are talking here not of acceptable standards of living, which are envisaged as available for all members of a community, but of items desired as status symbols, by people who have perhaps no particular skill or experience that might make their request appropriate. A lone yachtsman, for instance, who had successfully sailed round the world, would fall in a different category in his need for a yacht.

Meantime my computer might advise me that high-powered cars were being discouraged in my area because of a shortage of fuel and the unsuitability of the terrain, and that the community recommended the use of smaller cars fitted with anti-pollution devices for local use, and land-rovers for the bush.

Two things come to mind regarding the fierce desire of both rich and poor for status symbols of this kind. Firstly, the prime concern of this investigation is not how to enable everyone to have a sea-going yacht for free, but how to enhance the quality of life of the very poor, and how to enable intermediate groups in society to maintain a reasonable standard of living without the psychologically maiming stresses pertaining today. For the time being, the rich do well enough, though I think many of them would gladly trade their wealth for a world free of terrorism, theft, pollution and atomic war. Luxury ranks rather low against fear.

Secondly, increased communication will widen our options, while increased knowledge and education open up new avenues of interest and delight unknown today. The shift in emphasis should thus diffuse some of these problems.

At a more mundane level, I might require a plumber to clear blocked drains or repair burst water pipes. If a plumber mends my pipes and we record it together on my terminal and on his, what we are recording is not the value of his service in terms of any form of monetary exchange, but the simple, essential fact that he has performed a useful service; that he is thereby contributing actively to society; and that he is therefore entitled to whatever may be his needs and desires at that moment, provided only that the system as a whole, or his immediate community or group, or some other individual, is able to supply them without unfavourable bias to the system.

As in natural systems, recompense for services rendered does not necessarily come direct from me, though it can do so. Nor is any useful purpose served by trying to put a value on what the plumber has done. However, too long a gap between such records of service might render him suspect for not carrying his weight, and would reduce the esteem in which he was held by his fellows. The same would of course be true for me, were I not performing a useful service in some other sphere.

Here it is necessary to stress once again that, in these hypothetical situations, what the computers are being programmed to do, is not to 'supply' or 'withhold' the goods or services required, but to correlate demand and supply by *putting people in touch with people*, in such a way as to maximize the capacities and satisfy the requirements of each of them. Nobody would be empowered to instruct another as to what his needs should be, nor in what manner he should contribute. Learning new skills would be considered as a valuable contribution to society, as would study, research, counselling or any other form of caring activity.

It could be argued that many people don't know what they want to do, nor, come to that, do they always know the potential of their own capacities and gifts.

In the first case, such people could put the question: Is there any need in our immediate community to which I, lacking specialized skills, could contribute? Alternatively: Are there facilities in our immediate community where I could learn a useful skill, and if so, what? Acting on data in its memory bank concerning such an applicant's background and history (assembled from data supplied by himself in the course of his life) and being in constant touch with facilities in the community, as well as with the immediate needs of all people and groups in the neighbourhood, the computer would have no difficulty in channelling such an applicant towards reciprocally satisfying activity.

In the second case computers could be programmed to conduct aptitude tests, or to direct the applicant to a suitable qualified person for the purpose.

In these and other ways, still to be worked out, *the efficient co-ordination of diversity* could ensure that, for instance, an old aboriginal passing on his knowledge of the wilderness, a paraplegic working from a wheel-chair, or an un-schooled woman caring for a child would be independent of charity and un-needing of greed. Each and all would become part of a true community.

In Servan-Schreiber's memorable words:

> Computerization is exponential. It is a vast, infinite tool at the service of man's imagination, ideas and dreams, a means of linking each person's individual work, whether routine or creative, to the whole. [1]

Because of these capacities, one can say without exaggeration that 'the mighty living system'[2] created by human activity can, with the help of computers, 'search out and find all the elements it needs and deliver them to the right place at the right time'.

CHAPTER THIRTY SEVEN

8.4 What a Computerized Living System is NOT

Summary: While a computerized society embodies neither an utopia nor any known economic system, it will make possible a situation in which many of the money-induced ills of society will fall away.

For the moment, then, the problems remain the same. What has changed is our ability to deal with them, combined with a new vision of social relationships in a living Earth.

Before going on to touch on possible means of change from the old to the new, I want to summarize some of the things which a computerized living system is NOT.

Here they are:

- A living network is not an Utopia, since people, with all their idiosyncrasies, remain just people.

- It is not a welfare state, since no-one gets anything for nothing. Everyone contributes something of value.

- It is not capitalism, since no-one charges, pays, levies rents or interest, makes profits or accumulates capital.

- It is not communism, since there is neither state-planned production nor party power. Nor is there revolution as a means to social justice.

- It is not a free-for-all anarchy in which no-one need do anything.

- It is not a system in which the lazy live on the industrious.

- It is not a system in which 'they' supply and 'we' receive, free of charge, without labour or input from ourselves.

- It is not a system in which the 'haves' live on the backs of the 'have-nots'; nor of the 'have-nots' relying on aid from the 'haves'. Nor is it permissible for the 'have-nots' to take from the 'haves' as of right.

- It is not a transfer of the world's wealth, for wealth is found to mean something quite different, and money no longer means anything.

- It is not an indemnification process.

- It is not a world of wasteful consumerism and litter.

- It is not a world of forced idleness and boredom.

- It is not a breeding ground for alienation, economically induced breakdown or suicide, though these things may still occur for other reasons.

- It is not a breeding ground for drug addiction, since pedlars will no longer reap profit from it.

- It is not a happy hunting-ground for poachers, since there is no longer money to be made.

- It is not a world of preparedness for war.

- It is not a world of pollution, misnamed progress.

- It will not leave millions of people dying of hunger, nor allow the malnutrition of untold numbers of children.

- It will not despoil, pollute or destroy the living Earth.

CHAPTER THIRTY EIGHT

8.5 Some 'BUTs' and 'WHAT IFs?'

Summary: Objections raised to the notion of oblique barter and a moneyless economy are seen to be invalid.

Here are some 'BUTs' and 'WHAT IFs?'

Three main objections to an increase in computer usage crop up consistently in different forms in the many works by people who either dislike or fear them.

The first is dehumanization and the loss of jobs. The second is intrusion into personal privacy. The third is the abuse of power.

We've already discussed the first at some length, and discovered that contrary to the horrors of an 'Orwellian' world, computers, correctly used, can in fact *re-humanize* what is at present a wholly inhuman and unacceptable situation for the bulk of the world's population.

The second, intrusion into personal privacy, depends, I think, on how each small group or community chooses to program its computers. This would be by common consent and might vary from group to group. While all knowledge and general data should always be available to all, some personal data might more suitably be kept confidential, to be released to a third party only on the authority of the card-holder when offering or applying for employment or other personal contact, or when requesting help. (Examples might be age, marital status, divorce, mental health, physical deficiencies, religion, qualification or experience, or where one lives - as for instance a woman living alone.) An exception might have to be made here in a case of serious crime.

At present the need for privacy is often linked to the monetary system. If this were still partly in operation, a banking account would naturally enjoy the same confidentiality which it does now. But if the monetary system itself disappears, this difficulty disappears with it.

I am sometimes told that you couldn't free all knowledge to everyone, either. 'Supposing every madman could make the atomic bomb?' In my view every madman already does. But what this kind of viewpoint overlooks is that making nuclear weapons is a vast and costly corporate effort. It can't be done by one man with a grudge against the world. Unless we all, by our corporate will, squander vital resources on a vast scale to build the reactors, no-one can press the lethal button. And if there are no millions to be made by building them, we may no longer find them necessary or wise.

The third main objection to widespread computer usage is its undoubted potential for the abuse of power. The argument goes like this:

> Even if what you suggest were possible, it would never be allowed. Vested interests, power blocks, giant corporations, and all the machinery of accumulated capital would see to it that the scheme did not work. Capital is power, and power is a basic drive in humankind. What is more likely to happen is that power blocks gain control of these sophisticated information techniques, and use them further to enhance their own control.

There's a very real danger here. It lies behind the widespread fear that Big Brother will be enabled to take over our lives and invade yet more of our cherished autonomy. This, without doubt, is what many groups, interests, and even governments are attempting to do. The way to confound them is so to enmesh information channels world-wide that it is no longer possible to keep data exclusive to any one group, for whatever purpose or under whatever name.

Whether the nation-state could survive such a system, or indeed whether it should, is open to debate. The point I want to make is that vested interest thrives on information someone else does not have, and maintains secrecy concerning its own involvement and intentions for as long as possible. The same applies to power, whether political, economic or military, to the point that there is no longer any way of separating these three. The issue is further confused by the money-motivated criminal element, whose power is already greatly enhanced by the ability to gain any unauthorized access to remote computers in which such intelligence rests.

This thriving activity, we are told, is getting rapidly out of control, since there is no code invented by man that cannot be cracked by man. The whole history of military intelligence and counter-intelligence rests on this fact. Moreover, the activities of giant computer-linked power-groups become increasingly difficult to disentangle, since whole packages of data can be switched from one computer to another across national boundaries without a final or national resting place. Ironically, this very fact, which favours at the present time both criminals and vested interests, will be self-defeating as the networks become further enmeshed, finally moving out of anyone's exclusive control - a self-regulating flow of electrons stemming from the corporate will of many minds, and beyond the power of *anyone* to envisage or manipulate.

The important point to remember here is that, though no single person or group of persons can control or manipulate a system so widely diffused, it will nevertheless be the first spontaneous, co-ordinated expression of mankind in the whole history of the planet. In it is embodied the needs, will, character and dreams of every separate human being on Earth. No single person, no group and no government yet devised has been able to accomplish such a feat. Yet it is no more impossible nor astounding than what we experience daily (without astonishment!) when we breathe, move, see, hear, or feel, or when a dragonfly hovers over the water.

All these simple acts embody miracles of co-ordination whose manifold and separate sequences are beyond our conscious control. Yet they occur, in health, with such harmonious and unobtrusive regularity that we barely notice them. There is no reason to suppose that a living global system would behave any differently. Only if an overwhelmingly large percentage of humankind were innately evil could the truthful synchronization of his spontaneous will be evil too, in which case the sooner we blow ourselves off the planet the better! Not for a moment do I believe that this is so.

Now to the 'BUTs'and 'WHAT IFs?'of money itself.

Money was for a long time the best tool available with which to regulate the flow of goods and services around the world. It is no longer the best. It has been outstripped by electronic tools with which it cannot even begin to compete. Yet we are so accustomed to using it, so delighted to own it, we balk at the very thought of giving it up, and unwittingly think up reasons why this cannot be done.

I am told that a moneyless economy would not work because the productive man would have his rewards ripped from him to pay the way of the non-producer/crippled/retarded, or whatever. One is tempted to ask: Ripped away by whom? By the government? In this scenario, the community is self-governed in so far as its economic affairs are concerned. The political connotations of economic self-government are of course enormously far-reaching, and the two functions are at the present time inseparable. Doubt has already been cast as to whether the nation-state as we know it today can survive the buffeting to which it is being subjected by a variety of powerful new trends. It is beyond the scope of this book to follow where such thoughts might lead. But supposing the nation-state did survive into the era of a self-governing moneyless economy, would the producer be forced to 'carry' the non-producer to any greater extent than he does today? Isn't this exactly what happens at the present time, either by direct taxation, or by 'charity' in its manifold forms?

When a man goes on the dole, is he not being carried by the rest of the community? When free or subsidized hospital services or health care are made available to the poor, does the money not come from taxing the more successful members of the community? When free or subsidized education is provided by the state, is it not, in the final analysis, the productive man who pays for it? Worst of all, are not enormous sums of money engulfed in administering it?

Under the present system, the situation is worsened by the tendency of the unemployed, the desperate, and the natural criminal to steal from the rich and the not-so-rich alike, both of whom may have contributed to the productive competence of the community by their initiative, their managerial skill, their brain power or their labour; and each of whom carries, in addition to direct personal loss and increased insurance premiums, part of the enormous administrative costs of police, prisons, and courts. I do not see how removal of the money motive could fail to reduce the colossal waste of such unproductive activities.

Then there is the question of 'aid' to impoverished nations who, for a variety of reasons, are unable to feed or care for themselves. Because they have no money to pay for the excess food stock-piled in other parts of the world, nor the infrastructure to distribute it, millions of people die. Alternatively the 'rich' nations lend these peoples money with which to buy the products they need, knowing full well that they have not the slightest hope of ever paying it back, or even of keeping up interest payment over an extended period. Once again these moneys come out of the pockets of the productive section of the population.

To make things worse still, these moneys do not always further the projects for which they were intended. One definition of 'aid' that went the rounds recently was 'taking from the poor of the rich countries to give to the rich of the poor countries'!

Finally the situation gets so desperate that food is rushed to the scene by caring volunteers and governments alike, without expectation of payment of any kind, and once again it may or may not reach its destination. Lack of infrastructure, wars, rebellions and just plain incompetence all stand in the way. Food rots on railway sidings and people die. Would it not have been better to circulate excess food in the first place, while taking basic steps to help people help themselves? A moneyless economy on the lines of a living organism would enable the world to do just this, without all the preliminaries, and people with food in their bellies and hope in their hearts would be in a far better position to start producing for themselves.

I am told that about a fifth of the world's population is psychotic - motivated by the continuous urge to harm or destroy; that the damage they can do is out of all proportion to their numbers; that unless they can be detected and isolated they will:

- wreck the proposed system by cutting or distorting the communication lines

- see to it that those least deserving of rewards get most of them

- enter false data into the network system.

I am told that this will make it very tough on the remaining 80% of decent people, who will not know where the trouble lies nor what to do about it.

I have no data on these matters, but supposing we take them at face value as they were presented to me. Once again one might ask: Do these things not happen now? Do power lines not get cut by terrorists? Do lines of communication not get sabotaged? Do trains not get derailed nor tankers sunk? Are bombs not planted in planes with frightening frequency? Is massive electronic crime not already a major headache to the authorities, and are not ordinary decent men and women helpless to do anything about it?

As things are today, these things are done not just by psychotics, but by those who see profit in it, as well as by politically motivated fanatics. Since politics, too, are so often economically motivated, I find it difficult to understand why the situation should be worsened when the monetary motive is removed.

Prigogine's observations on diffusion in living systems (p.17) are very pertinent here. The more complex a system is, the more numerous are the types of fluctuations that threaten its stability. Part of the answer, according to Prigogine, lies in the diffusion processes of communication, the very processes we are trying to promote.

Without doubt the psychopath with an urge to destroy will apply his cunning and skill to this end. But where will he start? He can wreak havoc locally, to be sure, and cause untold inconvenience and misery to a great number of people, just as he does now. But in a system as widely diffused and enmeshed as is envisaged here, he cannot disrupt a significant part of the system, let alone the whole of it. The day-to-day ordinary activity of the 'decent 80%' will keep the thing in balance, and that far more effectively than they are able to do at the present time.

As regards the mentally or physically handicapped, the aged, 'the lame, the halt and the blind', are they not carried by the productive section of society to a far greater extent at the present time than they would need to be when their value was no longer assessed in monetary terms? And what about those service-oriented men and women whose satisfaction lies in voluntary service to their fellow men? Would such people suddenly lose their sense of dedication because no money was involved? I cannot see it that way.

Must we believe that the farmer, whose pleasure it is to cultivate the land, will cease from following his chosen way of life because he no longer receives for his products money he no longer needs? that the educationalist will lose his interest in the young, or the golfer his love of golf?

In the self-regulated economy envisaged in this book, everyone will be enabled to follow his bent as never before. Everyone and no-one will be under-privileged. Everyone and no-one will be privileged either. The productive man, the creative man, the dreamer and the genius will work, and dream, and struggle, and achieve, just as he does now, but much more creatively, much more happily, for he will not suffer the crippling distress of feeling himself better off then his fellows, nor stagger under the weight of carrying them.

Book III - Part Nine

Steps Towards Transition

CHAPTER THIRTY NINE

9.1 The Business World

Summary: Only extreme flexibility of approach can use the chaos of today's business world as a positive factor in social integration.

All innovative action starts with an idea. The innovative society, like the innovative company, is no exception. Drucker suggests that ideas are like babies - they are born small, immature and shapeless. They are promise rather than fulfillment.

> In the innovative company executives do not say, 'This is a damn-fool idea'. Instead they ask, 'What would be needed to make this embryonic, half-baked, foolish idea into something that makes sense, that is feasible, that is an opportunity for us?[1]

To say that my ideas on an organic economy as presented in *The Cosmic Microbe* are innovative in the extreme would still be an understatement. And of course not all ideas are sound. Drucker suggests that innovative ideas are like frogs'eggs:

> Of a thousand hatched, only one or two survive to maturity. Executives in innovative organizations therefore demand that people with ideas think through the work needed to turn an idea into a product, a process, a business, or a technology.[2]

Since these activities are typical of modern man, the required criteria can with justice be applied to the wider issues of society, and indeed this is exactly what I shall now try to do in the final stages of this book: The issue is this:

What work would we have to do and what would we have to find out and learn before an organic, self-regulating economy could come into being?

To begin with we have to realize that the position in which we find ourselves is totally different from anything there has been in the past. For this reason it is no use trying to go backwards, however deep our longing for stability or our nostalgia for the 'simple' life of our ancestors. Even should we desire it, which most people do not, the hunter-gatherer life is closed to the majority of us for ever. And though everything we do and are is part of a continuum embodying the whole of mankind, past and present, the road out of chaos is a forward road without historical precedent to mark out the way.

The next thing we have to realize is that the stability in which one could confidently predict the future ten years, or five years, or even one year ahead simply does not exist any more. The economic situation is totally unstable and unpredictable, and requires entirely new tactics to keep it operative.

Thirdly there is the fact that effective national control of an economy is a thing of the past. Not only is any and every nation, big or small, vitally and increasingly affected by events elsewhere on the globe, but despite all efforts at protectionism, crossborder transactions in manufactures and trade grow daily more involved. Finished goods, components and subcomponents are increasingly made and imported across national borders,to the point that components of even so proudly American an article as the IBM personal computer have at times been made abroad.

It used to be accepted that capital movement followed the demand for goods and services. That has now changed. Drucker points out that in the last ten to fifteen years capital movement as such has become the engine and driving force of the world economy.[3] Could anything more ridiculous be imagined? That the production of day to day needs of the populace should be dependent on anything as arbitrary as capital flow and the whims of the capital market smacks of insanity even in today's zany world.

Capital movement today is essentially international. It follows that today's economy is essentially global in character and can no longer be effectively controlled by governments, which makes nonsense of national macro-economic theory. In so far as the economy is concerned, it also makes a laughing stock of the nation state.

The situation is further exacerbated by the games people play with money in the field of currency exchange - serious and sometimes deadly games that work havoc with the lives of countless millions of ordinary people, but games none the less. In the international monetary world nothing is predictable from one day's financial news to the next. Currencies swing wildly by as much as 5 per cent in a week, resulting in currency speculation and manipulation that turns the economy of whole nations completely upside down. When we read of currency-exchange transactions totalling $80 trillion a year, of which only $4 trillion is required to finance trade in goods and services,[4] the absurdity of the whole tragic situation passes belief.

Add to all this that the planet itself is no longer able to deliver the air, water, climate and food that humanity needs unless we drastically and quickly mend our ways, and it becomes clear that the situation has passed the point where tinkering with an obsolete financial system, even on an international level, is adequate. We *have* to think our way through to something new. It is at this point that a self-regulating, organic and living system begins to make sense.

In his book *Thriving on Chaos*,[5] Tom Peters asserts that in the modern business world chaos is now the norm; that everywhere and every day managers confront shattering and accelerating change; change paced by constant innovations in computer and telecommunication technologies. Realization that these innovations might also hold within them answers to the chaos they have partly caused, seems not yet to have surfaced in the business world. But my own assessment is in line with Peters when he presents chaos as the very foundation for future progress:

> To survive and become superlative in today's economic environment there is an urgent imperative beyond excellence: the flexibility that can - and must - use chaos as the source of market advantage.[6]

I would differ from Peters only in one respect, but it is an important one. What the world economy needs is not market advantage, but market *integration*.

It is illuminating at this point to draw comparisons between Tom Peters' book *Thriving on Chaos* and that of the Belgian chemist Ilya Prigogine, *Order Out of Chaos* discussed earlier. It would seem that in business, as in dissipative chemical systems, the organization will be retained as long as fluctuations remain within certain limits. Beyond these limits, they will drive the system into instability. This moment of extreme instability creates the crucial conditions for evolutionary change. The system will either collapse or attain a new level of organization.

In business as in natural systems, the more complex a system is, the more numerous are the types of fluctuation that threaten its stability. To explain the astonishing fact that systems as complex as ecological or living systems manage to exist at all, Prigogine suggests that the two principles of diffusion and communication are likely to be significant - diffusion through the interaction of many species and individuals, and communication amongst the various parts of the system. Once again it is significant that at this present moment of chaos in human economic affairs, partly brought about by the rapid development of computers and telecommunications, it should be diffusion through communication that offers the best route to an acceptable threshold of stability. I shall argue that lively interaction between large numbers of individuals, on a scale not previously seen, has been made possible by electronic technology, and that this interaction will not only provide a stabilizing effect on the economy of the future, but will release the innate creativity of the individual while integrating him closely into society. This time it will be people, not capital, who provide the motor power of the network - people who are rapidly becoming aware of the dangers we face and the personal responsibility each one of us bears in meeting them. We are already beginning to act on this in our encouragement of small business enterprise, as well as in public demand for an environmental ethic. Individual investors and consumers already exert considerable pressure on major profit making organizations, and this trend grows yearly more vociferous.

Any movement towards ecologically sound business practice will have to take into account that man is a major component of the environmental system in which he operates. Because of this, any activity within the framework of a self-regulated, organic, non-monetary economic system such as I have been trying to describe will have to take cognizance concurrently of both man and his surroundings, and of the effect of each upon the other. Without this his efforts will be doomed to failure. Nor will passionate concern and action be enough. In moving towards the systemic approach in economic affairs, we must do our utmost to trace the factual connections that will relate what we do to what we get. This does not, of course, mean that what we get in the long term will be exactly what we envisaged. It has already been noted that in a complex system a single variable will throw a whole matrix out of balance, and that this is particularly true of living systems. To keep control during the transition we may have to speed up some developments and slow down others, but because it is the quality of the individual mind that generates the nascent collective, what we get in the end will reflect faithfully what we are.

The Way Ahead

How do we get there?

The way to figure out what has to be done, says Negroponte, of the Massachusetts Institute of Technology, 'is through exploring the human sensory and cognitive system and the ways that humans most naturally interact. Join these and you grasp the future.'[7] Negroponte is talking about development in fifth generation computers, but in economics as well as in computer science, I believe this is the way to go.

How do human beings most naturally interact? In small groups of people who know each other and share a common interest, a common environment, a common need or a common love. So the economy of the future will link such groups in local cycles of production, usage and services based on information concerning each other's need, and the exchange will spread in ever widening and changing patterns through a network of information for which we already have the necessary technical competence. We have already spoken of the capacity of current technology to bring together people who would otherwise have no means of getting to know one another. What we shall have, therefore, is a fluid but closely knit system that faithfully reflects the personal reaction of individual people, but does so increasingly on an ecological basis rather than on a national one. In economics as in cellular health and balance, it will all be a matter of having concerned neighbours who keep an eye on you.

I am often told, when discussing these things, that self-regulation is a recipe for anarchy. It would indeed be strange if this were so. We know that random events, provided there are enough of them, produce patterns with reason and meaning, and that this is true throughout the known universe, right down to the level of sub-atomic events. It has to do with the mathematical law known as Bernoulli's law, or the 'Law of Large Numbers'.

Says Daniel Hillis of MIT's Media Lab:

> What makes physics look beautiful, so you can write down 'f=ma' (force equals mass times acceleration) - those lovely linear laws? It's not because anything simple is going on at the bottom level. It's because when you have the cumulative behaviour of billions of things together, then what you can describe is the gross trends of the system, the net effect of all that mass. It's the net effect that's beautiful and simple and elegant.[8]

When we finally get around to feeling ourselves intrinsic to the universe around us, and to the planet Earth in particular, we shall perhaps accept more easily that its laws apply to us too, and be less afraid of committing ourselves to patterns and processes that have worked throughout the millenia for all of nature.

During the build-up to the Gulf War, we were told that America was in financial crisis. The greatest economic power in the world was out of funds. We were told that its debt had reached the staggering figure of $350 trillion, and that the President's budget aimed at reducing this figure had been rejected. To keep the legislature going at all, a ten-day moratorium was declared, while economic experts tried to reach a compromise, but so violently opposed were the views of conflicting parties that one marvelled they could all be operating on the same set of facts. Meanwhile military might built up in the Gulf, the price of oil soared, and we teetered on the brink of a Third World War in which everyone knew there could be no winners. It came about all the same. And so interlocked are the economies of nation states that banks and large corporations in effect funded both sides, while the military might of Iraq was partly composed of arms supplied by her current enemies. So interlocked, indeed is this global involvement that no nation is exempt from repercussions when another collapses, and the Allies now have to build again what has so lately been destroyed.

Ironically enough, concerned bankers of the 'Big Seven'regularly put their heads together even in peace-time to bolster the U.S. giant, lest the whole pack of cards collapse.

No-one knows what the outcome will be. There is talk of hard landings, soft landings, and no landings at all, with a variety of variables making it impossible to predict the course of events even from day to day. The so-called 'symbol economy' of international currency and capital flow seems very much in command, with fortunes made and lost in ways having little to do with the bread and butter issues of the ordinary world.

Considering what it can do to the life of the small man and to the less developed nations, this kind of gamesmanship in world finance seems almost malign, yet I think it is not deliberately so. It is rather that, in placing too much emphasis on profit as the flywheel of growth and development, we have created a situation uncannily like the fabled Polyploid Horse. Gregory Bateson tells how, in the 1980s, the great 'Erewhonian geneticist' got a Nobel Prize by jiggling with the DNA of the common cart-horse to produce an animal twice the size of the ordinary Clydesdale.[9]

It was twice as long, twice as high, and twice as thick. It was a polyploid with four times the usual number of chromosomes. But by the time this poor animal reached maturity it had more than its share of problems. *It was too heavy to stand.* It weighed, of course, eight times as much as a normal Clydesdale.

> For a public showing and for the media, Dr Posif always insisted on turning off the hoses that were continuously necessary to keep the beast at normal mammalian temperature. But we were always afraid that the innermost parts would begin to cook. After all, the poor beast's skin and dermal fat were twice as thick as normal, and its surface area was only four times that of a normal horse, so it didn't cool properly.
>
> Every morning, the horse had to be raised to its feet with the aid of a small crane hung in a sort of box on wheels, in which it was suspended on springs, adjusted to take half its weight off its legs.
>
> Dr Posif used to claim that the animal was outstandingly intelligent. It had, of course, eight times as much brain (by weight) as any other horse, but I never could see that it was concerned with any questions more complex than those which interest other horses. It had very little free time, what with one thing and another - always panting, partly to keep cool and partly to oxygenate its eight-times body. Its windpipe, after all, had only four times the normal area of cross section.
>
> And then there was eating. Somehow it had to eat, every day, eight times the amount that would satisfy a normal horse and had to push all that food down an oesophagus only four times the calibre of the normal. The blood vessels, too, were reduced in relative size, and this made circulation more difficult and put extra strain on the heart.
>
> A sad beast.

Bateson uses the fable to show what inevitably happens when two or more variables, whose curves are discrepant, interact. The changing of one variable exposes a critical value of the other.

In the case of the imaginary horse, its length, surface area, and volume (or mass) become discrepant because their curves of increase have mutually nonlinear characteristics. Surface varies as the square of length, volume varies as the cube of length, and surface varies as the 2/3 power of volume.

For the horse (and for all real creatures), the matter becomes more serious because to remain alive, many internal motions must be maintained. There is an internal logistics of blood, food, oxygen, and excretory products and a logistics of information in the form of neural and hormonal messages.

In economics, as in living things, this involves literally millions of variables, which is why attempts to model possible scenarios so often fail. But I think it is fair to say that overemphasis on the profit motive, widely held to be the motivating force of the free market system, has thrown everything else out of balance and rendered the system all but unworkable.

Capital and the Profit Motive

We in the Western World are rather smug just now. Things are happening that would have seemed impossible only a few years ago, like the end of the cold war, the collapse of communism in Eastern Europe, the re-unification of Germany and a general movement toward a free-market system. Private enterprise and its partner, the making of profit, are regarded in the West, and increasingly in the East, as the motor that drives the whole economic enterprise. We've been pretty successful too, if you discount small anomalies such as that most of the world's population lives below the breadline, that most human activity doesn't figure in the economic equation, and that the world is on the brink of ecological disaster.

Man has three basic material needs for survival: clean air to breathe; pure water and food to eat; shelter from the elements. Beyond these he has a need for useful and satisfying activity to nourish body and mind, and a transcendental goal to feed his spirit. Unless capital- flow and the making of profit can meet these needs for the vast bulk of the human population, they have no place as the driving force of anything at all.

When I was a child I was told: The poor will always be with us. And this comfortable assumption, usually by those who had no experience of poverty, cropped up at regular intervals to reinforce the notion that poor people were somehow outside the area of our responsibility. They were almost, it seemed, a law of nature, like rain when you wanted a picnic, and while it was one's duty to be kind or charitable to the stranger at one's gate, it simply did not dawn on us that it was possible to reorganize human affairs or change the weather. Today, with knowledge more freely available than ever before, continued passivity is simply not on.

We've seen enough of the effects of systems and ideologies imposed from the top to know that they do not work, and the collapse of Eastern Bloc economies in recent years only goes to re-enforce this view. But we cannot afford to be smug while untold millions of people live and die in misery. 'The simple, and terrible, truth is that poverty and environment are inextricably linked in a chain of cause and effect. Problems of environment cannot be tackled in isolation from those national and international economic factors that perpetuate poverty.'[10] Nor can we allow the pursuit of profit to pollute the skies, engender the greenhouse effect, destroy the ozone layer, play havoc with the world's climate, destroy the fertility of the soil, decimate the rain-forests or pollute the seas.

So neither system will do. It is as simple, and as difficult as that. And we haven't much time. If humanity is to survive the turn of the century, we have to be brave enough to question obsolete ideas and really begin to think. There are encouraging signs that large numbers of highly competent and dedicated people are doing just that.

CHAPTER FORTY

9.2 Natural Principles in Economics

a.) Industry
b.) Household Wastes
c.) Agriculture
d.) Economic Viability

Summary: There are three essential ways in which natural living systems differ in their operation from the man-made economic criteria on which we operate today.

Natural Principles in Economics

There are three essential ways in which natural living systems differ in their operation from the man-made economic criteria on which we operate today.

The first is that in a living system individual units, while pursuing their own concerns, unconsciously contribute benefit to some part of the system both in their living and in their dying, but this benefit is not usually in the form of a quid pro quo exchange. It is not usually direct. This is the process I have referred to earlier in this book as oblique barter, and would include, for man, those intellectual and spiritual intangibles he is pleased to think of as essentially human attributes. The notion of direct payment for services rendered or goods received as a prerequisite for such services simply does not occur in an organism, nor, for that matter, in nature.

The second is that, in an organism, processes are not organized around energy as defined by the physicist. They are organized by information. And in the healthy organism all information moves freely in an uninterrupted flow throughout the system at all times.

The third is that, within the system itself, there is neither waste nor pollution. No substance is secreted or otherwise produced which is going to prove harmful and even lethal to other parts of the system and to itself. The marvellously intricate chemistry of any living body operates to sustain itself in health. It is able to do this because of the universality of its information system, whereby the life of every individual cell and every organ is integrated in the life of every other.

If we regard the Earth as a living organism, it is very clear that at the present time we are not fulfilling these requirements. We are violating the organic principles evident elsewhere in nature, both in over-exploitation of resources and in the production of dangerous quantities of toxic waste. There is no waste in nature. There is redundancy, but this is another story. All products, wastes, substances and creatures in nature return ultimately to the air, the soil or the sea, and are reabsorbed into the cyclic system. In other words, they are bio-degradable. There is no natural process which, within the natural lifetime of the planet, threatens the existence of the whole living system.

Man, if he and the planet are to survive, must embrace these three principles, which are complementary and inseparable one from another. The first he will fulfill by pursuing some useful activity that both fulfills his own nature and directly or indirectly benefits someone or something else.

The second he will achieve by diffusing information, a process which is coming about with lightning speed as computers, satellites and telecommunications begin to encircle the globe.

The third he will fulfil by ceasing to manufacture or produce anything that cannot be absorbed or used as a resource by some other part of the system. He will be enabled to do this because of the ubiquity of the integrative information system, which vested interests and power groups will increasingly find it impossible to control, and by which, eventually, all research and knowledge will be freely open to him on a day to day basis.

Waste As A Resource

a.) Industry.

I want, first, to talk about this vital issue of 'waste as a resource', because it is crucial to the survival and health of the planet, without which we need not bother about economics or even about man himself. Also because although there are enormously hopeful trends in the emergence of a new breed of 'Green Capitalists,'[1] the efforts of these pioneers have until recently been largely confined to dealing safely with toxic wastes prior to disposal. With a few notable exceptions they have not yet, so far as I know, made it general company policy to establish a profitable end use for the detoxified product in a subsidiary industry. In this they have stopped short of the way of natural planetary processes, in which all substances are part of the cyclic flow.

Dumping

Part of industry's answer to the disposal of wastes has been to dump it in 'safe' localities in the sea. Such practices are anything but safe, for the effects are cumulative. But because the sea is vast, it has taken a long time for the results of our foolishness to catch up with us. The dangers are now well known. The practice continues. Though international attempts to legislate have been under way for some time, their success has been limited, for without viable alternatives, the drive for profit ensures that the dumping goes on. Gourlay observes that 'with more than half a century of agreement on what needs to be done, it is time for the lawyer to hand over to the policeman and ensure that it *is* done'.[2] I don't think the answer lies this way. Economic problems in industrialized countries ensure that governments drag their feet, while the Third World, desperately needing development, is openly hostile to restrictive laws. Moreover, whatever the legislation, it is almost impossible to enforce.

Perhaps worst of all, and this time inexcusable, is the current practice of shipping dangerous wastes, prohibited in industrialized countries, to the Third World, whose need for foreign currency often ensures their acceptance.

The problems are global, and cannot rightly be laid at anyone's door. But we do now know that they cannot go on. People are poisoned by toxic fish. Marine life on which we depend is being destroyed. The oceans are sickening and the sea will die. We *have* to find new ways. It is as simple as that.

For those unfamiliar with the extent of the damage, perhaps a few examples will suffice.

Research on pollution of the North Sea revealed the following figures in 1970: that the Dutch poured in 3,600 tons of sulphuric acid and 750,000 tons of sulphur dioxide each year; the West Germans dumped 375 tons of sulphuric acid and 750 tons of iron sulphate daily, 40 tons of chlorinated hydrocarbons and 40-50 tons of polyethylene every month, and 200,000 tons of gypsum every year.[3]

These were the official figures. Given the reluctance of people and firms to admit responsibility, together with the widespread practice of clandestine dumping, actual figures will surely have been higher.

That was twenty years ago. The full effects haven't hit us yet. Though fishing grounds have been severely affected, we do not yet know fully what the repercussions will be. A glance at the next example may help us guess.

Minamata

Minamata was, in the 1950s, a town of some 50,000 inhabitants on the east coast of Japan. Many of its inhabitants were fishermen, though the town's economic prosperity depended on the Shin-Nihon Chisso Hiryo chemical factory. Among the waste chemicals discharged into Minamata Bay over the thirty years up to 1968 was the highly toxic material methylmercury, but it was fifteen years before, in 1953, the first signs of disaster appeared. This took the form of a mysterious disease, later to be known as 'Minamata Disease,'affecting members of the fishing community and their families. Three years later it had become epidemic.

The illness began with numbness in the lips and limbs. After two weeks the patient's senses of touch, speech and hearing were impaired; his gait became irregular, he walked as if drunk, and he was unable to stop suddenly or to turn around. More disturbingly still, the field of his vision was narrowed. The expected end would be madness, death or permanent disability.

Professor Shimanosuke Katsuki of Kuomoto University inferred from the symptoms that the cause of the illness was heavy metal poisoning, but though the staple diet of all the victims was fish, it was not until three years later that the source of the poison was established. Even then the chemical company refused to give information about its production processes or to submit samples of its waste for analysis. When, in 1960, investigation showed that the discharge contained methylmercury, they still denied all responsibility. A business-oriented government took no action, and only in 1969, sixteen years after the disease first appeared, was it was proved beyond doubt that mercury in organic compounds had caused the poisoning.[4]

By 1980 the lesson had still to be learnt.

The Kalu River

In the early 1980s the Institute of Science in Bombay revealed that industries on the banks of the Kalu River on the outskirts of Bombay were creating the conditions for another disaster.

'The spectre of Minamata stares in the face of the village population that consumes fish and food from or near the Kalu river,' declared the leader of the investigating team, Dr B.C. Haldar.[5]

The Kalu is a small river which flows from the Western Ghats through the industrial suburbs of north-east Bombay to reach the sea 50 km north of the city. During its course it receives effluent from over 150 small industries; the ten-kilometre stretch between Ambivali and Titwala, which was examined in detail, received toxic waste containing mercury, lead, copper, cadmium, chloride, dyes and organic acids from a rayon factory, paper mills, a dye factory and a chemical plant. Dr Haldar was in no doubt of the potential consequences.[6]

> The water at the point of discharge into the Kalu river has a mercury content equal to that at the centre of Minamata Bay.

Gourlay observes that here (as in most places) it is the industrial lobby, not the environmentalists, who decide events. 'Senior officials of the Maharashtra Board for Prevention and Control of Water Pollution have reputedly resigned after being frustrated by politicians in their attempts to ensure that industries along the banks of the Kalu conform to existing regulations.'[7]

In the 1980s and 90s we've been learning fast - albeit at the eleventh hour - that the only way to save our own skins is to save the living systems of planet Earth - including the seas and the skies - and that the only way to individual wellbeing lies within the wellbeing of those systems and of each other. The extent of popular concern in these matters is evident in the explosion of environmental issues into headline news, and the growing realization that confrontation between environmentalists and industry must give way to a balance, and even to cooperation between the two. International institutions such as the World Bank and the International Monetary Fund, staunch supporters of the free market system, are increasingly refusing funds to projects deemed environmentally damaging, and major projects have continually to be reformulated to meet stricter environmental standards. And as recently as April 1981 a group of five of the world's biggest corporations, including Shell, Mobil, and ICI is reported to have met in Holland to discuss (and we hope to tackle!) environmental hazards. Even more significant is the spread of concern amongst the general public in what is being called 'ethical investment'. A recent poll in the U.K. is reported to have found that of the people approached, more than fifty percent favoured high standards of environmental ethics in the companies in which they would wish to invest. Thus companies are being subjected to more and more pressure to reconcile their profit-making obligation to shareholders with the demand of these same shareholders for environmental acceptability.

Francis Kinsman tells how the sage and futurist Willis Harman recently 'wowed' a Business Network audience with the suggestion that business could end up as the prime mover in the great holistic paradigm shift that is now under way.[8]

Under the slogan 'there is no solution short of reinventing the world', he focused on four main pressure points - the hardest to deal with but the most essential:

- redefining development - the need for a totally new economic system embodying a range of economic cultures

- redefining links between people and society - the new nature of work and employment

- redefining national security - so that it no longer depends on military strength

- redefining science - so that it encompasses the complementary, the participative and the subjective.

In this endeavour, Harman argues, business has a vital part to play. It has the characteristics of flexibility, capability, sensitivity to change and clearly defined performance criteria.

I would go along with this. And if business parameters have moved in recent years - as they are clearly beginning to do - to include environmental excellence as well as profits, this can only be good. However, while some companies are on the right track, what does not yet seem to be generally realized is that *there is an alternative* to the suicidal accumulation and dumping of toxic waste.

While industries can and must reduce environmental hazards by spending large amounts of money on the cleaning-up process, the disposal of poisonous chemicals by discharging them into lakes and streams, dumping them in the sea, or burying them in 'safe' locations on the land, quite simply has to STOP. This means that *we have to find ways of using the waste products of each and every industry as raw material in another,* thereby satisfying environmental requirements, improving human and animal health, increasing profits and opening up new markets. By doing this we could begin to emulate natural processes, in which the waste product of every species is the nutriment of another, and the interaction of all species ensures the continuity of life.

Clearly, then, we are giving our engineers and chemists the wrong brief. What we should be saying is not 'Get rid of this waste with as little cost to the company as possible', but 'Find us a complementary industry in which we can increase our profit by using this waste as its raw material.' Alternatively they could be asked to separate dangerous wastes into useful components, or combine them in less harmful forms. This too could entail their use in a subsidiary of the same company, or their sale as a raw material for the industry of another. Either way they should where possible be used locally, either in the raw state, or after treatment.

If this sounds a tall order I can only say that the pursuit of profit at the expense of another man's livelihood or health, and eventually at the expense of humanity's resources as a whole, is both unethical and unsound. Given the will, the right brief, and the right money, chemists do have the expertise to transform dangerous substances into useful ones. Nothing - absolutely nothing - should be regarded as waste, to be 'got rid of', with all the attendant hazards entailed.

An example from ICI will illustrate what I mean.

Cyanide is what John Elkington has called 'a peculiarly nasty chemical'.[9]

> Swallow as little as 300 milligrams (just one hundredth of an ounce) of cyanide salts or inhale a mere 100 milligrams of hydrogen cyanide gas and this rapidly acting poison will disrupt the fine chemistry of your cells. Breathing air containing as little as 200 to 500 parts per million of hydrogen cyanide can prove fatal within half an hour. However, hydrogen cyanide is also an extremely useful industrial chemical, being used in the manufacture of acrylic fibres, synthetic rubber and plastics. The uses of cyanide have extended well beyond the manufacture of the synthetic fibres which are now such a feature of modern life. They include the manufacture of synthetic dyes and the hardening of iron and steel.

Bearing in mind the dangers involved, we learn with revulsion that cyanide residues have nevertheless been consistently and indiscriminately dumped throughout the developed world.

In 1982 an ICI research team based at Billingham started work on an ingenious process by which cyanide degrading organisms isolated from a fungus could be fermented commercially. Their research was based on the work of Professor Christopher Knowles, head of microbial biochemistry at the University of Kent:

> 'It's an elegant process and a simple one,' said Professor Knowles. 'A range of natural organisms produce cyanide, while others are able to degrade. 20 per cent of all plants produce cyanide, for example, using it as a defensive mechanism to ward off micro-organisms which would otherwise invade damaged leaves. We knew some fungi and bacteria seemed to be able to get around this defence, by breaking down the cyanide. Once we had realized that, we began to wonder whether

there might not be some commercial applications of this natural detoxification process.[10]

The work of the Billingham team was so successful that the potential of its enzyme product, 'Cyclear' in the international market was estimated to run into tens of millions of dollars in North America, Europe and Japan. Even more significant from an environmental point of view was its influence in triggering ICI's aspirations to use a wider range of biotechnology in the management of effluents arising from oil, petrochemical and chemical industries.

All this is extremely valuable, but one more step is required to bring it into line with natural methods. This step involves the use of all waste subsidiaries of the detoxification process as raw material in some other industry. Our chemists and our engineers must set their sights not only on the detoxification of wastes, but on their reabsorption into the industrial cycle. I hope to show that the same principles apply both to household wastes and to agricultural practices, and that while contributing to human and environmental well-being, their adoption could render unnecessary the economic rat-race in which industry is ensnared at the present time.

Crucial to an understanding of these principles is the perception that we ourselves are ecosystems. So we are dealing with an internal and an external environment, and the two are interlinked. It is in the interaction between the two that the health of both depends.

To illustrate the point, there's an interesting parallel between ICI's cyanide research and the work of Dubos on viral pneumonia.[11] Both are based on the perception that since everything in nature is destructible and recyclable, there will always be a microbe, a bacteria or an enzyme that can break down the most obstinate capsule in which some other organism may encase itself. Working on this premise ICI was able to find a bacteria - producing an enzyme that would break down the cyanide, and Dubos was able to beat the pneumonia virus. Taking this work a little further Dubos introduced the idea that disease-producing organisms are not inherently virulent. They are ecological in the sense that they are part of the total harmony of nature and of organisms. The whole thing adds up to interaction between ecosystems, both inner and outer environments, and in all healthy situations the thing that connects them is an all pervasive information system, neural, hormonal or computerized. The present surge of environmental awareness marks the beginnings of our understanding of this relatedness at all levels, including industry. As such it could trigger a shift in priorities in the human economic equation.

b.) Household Wastes

Moving to the field of household wastes it is fair to say that once again we are giving our engineers the wrong brief. In the densely populated countries of Mediterranean Europe, for instance, the problem of waste disposal had reached a point where the ocean was little short of a sewer, and marine life all but destroyed before serious efforts were made to clean up the mess. The same applied to the river Thames, and to countries bordering the Rhine.

Even today, after intensive cleaning-up programmes, you can walk along the sea-front of the famous Cote d'Azur of the French Riviera and search in vain for a live shell-fish or anemone in the pools. And on that most beautiful coast I have known the stench to be so intense that one gave up walking anyway.

Taking Holland as an example, there were, prior to the seventies, only enough sewage works in Holland to cope with about thirty per cent of the sewage; the rest went straight into the canals, rivers and lakes.[12] It is a terrifying thought that the same kind of situation still pertains today in many parts of the world, as witness the thousands of deaths from cholera in our own decade.

Kuenen remarks that because of their geographical position, the Dutch are particularly conscious of the international nature of European environmental problems. Above all, they are on the receiving end of the Rhine, central Europe's major river, and still its main industrial drain. The international character of this arterial river has made it immensely difficult to get anyone to take responsibility for its pollution, or to enforce the necessary legislation. It was always just too easy to pass the buck. Moreover the International Rhine Commission, established in 1950, had still, in the early 1970s, only got engineers on it - no biologists or environmental people. Kuenen observes dryly that water engineers know about microbial aspects and bacteria, but not about fish,[13] and indeed it is a tragic aftermath of twentieth century mechanistic thinking that we have been able to compartmentalize our activities in this way.

To pour untreated sewage into rivers, or even to regard it as waste for disposal is such a travesty of sensible procedure that it is difficult to believe intelligent men could permit it for so long.

In their excellent recent book *Blueprint for a Green Planet*[14] John Seymour and Herbert Girardet discuss in detail the relatively simple way in which this problem maybe approached:

What, they ask, happens to waterborne waste?

> In a sewage treatment plant the solids are allowed to settle out and the liquids - after bacterial action - are pumped out into the nearest river ... The solids, still fairly obnoxious after a drying process, are then either dumped in a landfill site, or, more and more, taken out to sea. Incineration has been tried but is expensive: sea-dumping is the preferred method wherever practicable. To take England and Wales as an example, very roughly one-fifth of the sewage sludge produced is dumped at sea, two-fifths dumped in landfill sites and two-fifths applied as fertilizer to agricultural land.[15]

Now all natural processes are cyclical in nature, while except where sludge is returned to the land, the process just described is entirely linear. The finite supply of essential plant nutrients - nitrates, phosphates and potash - passes from the earth, through plants and animals to the food-stuffs we consume. Thereafter the sludge is dumped, in either solid or liquid state, in our rivers and in the sea. Having thus discarded the precious nutriments on which agriculture depends, the final irony is the subsequent use of vast quantities of oil and natural gas in the manufacture of fertilizers to replace them.

As before, our engineers were given the wrong job. They should have been asked 'to recover for the use of agriculture the extremely valuable materials that occur in human sewage'.[16]

Surprisingly enough this is not, for developing countries at least, a difficult assignment. All that is required is for sewage to be put through a methane digester. Organic matter is put into an airtight tank in which anaerobic bacteria (bacteria which do not require air) get to work on it. In the process of digestion they produce methane gas which can be drawn off and put to use. The spent sludge, with the gas removed from it, is left as a completely safe, sterile and highly valuable fertilizer. Composting with low-grade organic material such as waste straw completes the simple, low-cost and invaluable process, recycling essential plant and animal nutrients, returning the land to good heart and contributing immeasurably thereby to human health. This is a perfect example of the reciprocal usage of 'waste' suggested above for emulation in industrial processes. In nature it is universal procedure.

The People's Republic of China is a good example of a country where this method is still in use. With one of the highest population densities in the world, China has preserved the fertility of its soil and supported its peasantry over thousands of years largely by returning human and animal excreta to the earth. We are told that here there is a methane plant in nearly every village, and they work perfectly. 'The village is provided with free lighting and heating, the land with a fine fertilizer, and what our engineers would call 'waste' is removed without any unpleasantness to anybody.'[17]

To switch to such a system in the highly industrialized urban societies of the world would seem to be a major, if necessary undertaking. 'Fortunately,' say Seymour and Girardet, 'the sewers we have now, in all the older industrial countries, are near the point of collapse. They will have to be renewed anyway. And when they are renewed we must insist that they are renewed in a completely different way.' While the word 'fortunately' might seem ironic, the message is clear. For developing countries it is clearer still: Healthy low-cost treatment of sewage in methane digesters is preferable in every way to the wasteful, high-cost, problem-engendering ways of urban society.

Seymour and Girardet sum the Western type situation up succinctly as follows:

> What we have done in Western countries is to make three problems out of the water cycle: the problems of how to fertilize the soil, how to prevent pollution of the rivers and seas and how to get rid of sewage. By mining out the stored nutrients of the Earth and, after using them once, dumping them into the sea where they are forever out of reach, we are robbing our posterity.[18]

We are also creating unnecessary costs. We need little money to return our wastes to the land.

c.) Agriculture

It must be forty years since I encountered, in Fairfield Osborne's pioneering book *Our Plundered Planet* (1948) the notion that something was gravely wrong in man's relationship with the planet on which he lived. It followed two outstanding studies: *The Living Soil*, by Lady Eve Balfour (Faber & Faber, London, 1943) and *An Agricultural Testament* by Sir Albert Howard (Oxford University Press, London, 1947), which later made so profound an impression on my young mind that I would never again be able to contemplate any of man's activities without wondering about their long-term impact on the environment. These early books were followed in quick succession by a rash of excellent publications on agricultural practice in different parts of the world, all of which not only highlighted the dangers of large scale monoculture and the indiscriminate use of artificial fertilizers and pesticides, but questioned their legitimacy in terms of human and animal health. They recognized, as we until recently have failed to do, that the real, permanent and vital capital of nations is the soil, and that failure to maintain its fertility by returning organic wastes to the land is to violate nature's essential processes and hasten our own demise. In agriculture, more than anywhere else, 'waste' is our most valuable resource.

> If mankind cannot devise and enforce ways of dealing with the earth, which will preserve the source of life, we must look forward to a time - remote it may be, yet clearly discernible - when our kind, having wasted its great inheritance, will fade from the earth because of the ruin it has accomplished.[19]

Those words were written nearly a hundred years ago by Professor N.S. Shaler of Harvard University. When Eve Balfour wrote her historic study in 1943 she observed that the future of which he spoke was no longer remote. Today we know that our methods in agriculture as well as in industry have destroyed our soil, our water, our atmosphere and our climate in a rapid lethal progression that some experts consider irreversible. We no longer wonder whether these dismal forebodings are valid, but how, in a world racked by poverty, wars and political upheaval, we can even begin to reverse their trend.

Public awareness of the dangers facing mankind seems to have peaked in the 1980's. In 1983 The World Commission on Environment and Development was set up by the United Nations to 'propose long-term environmental strategies for achieving sustainable development by the year 2000 and beyond'. Its report, known as the Brundtland Report, was published in London in 1987.

The Commission defined 'sustainable development' as ensuring that humanity 'meets the needs of the present without compromising the ability of future generations to meet their own needs'. It is 'a process of change in which the exploitation of resources, the direction of investments, the orientation of technological development, and institutional changes are made consistent with future as well as present needs. We do not pretend that the process is easy or straightforward.'

Meanwhile three new issues had hit the headline news: Global Warming due to the Greenhouse Effect, the Ozone Hole, and the horrors of a Nuclear Winter, all of which have been touched on earlier in this book. As though all this were not enough, nature had underlined our self-made plight in a series of droughts, floods, earthquakes and hurricanes that seemed to focus attention the more acutely on environmental concerns.

Said Shridath Ramphal, the Commonwealth Secretary-General, commenting in 1989 on the findings of the Brundlandt Report:

> The interrelated issues of environment and development now vie with nuclear disarmament as the dominant issue of our time. Politicians - from Mr Gorbachev to Mrs Thatcher - and financiers - from the President of the World Bank to environmentally 'clean' unit trust managers - advertise their 'green' credentials. But perhaps more important, was the way in which a succession of disasters all over the world triggered intellectual awareness about the possibility of some underlying pattern of causality, and aroused those emotions of fear and anger that are often the mainspring of political action.[20]

It was not before time. On the 1st May, 1991, came news of the most disastrous hurricane in recorded history. The storm which hit Bangladesh raised waves six metres high, submerging whole islands, whole villages, and all the surrounding countryside. Horrified, we watched our television screens. There was little left to see. Where only a few hours before, peasants had worked the fields, water stretched from one horizon to the other. Only the bloated bodies of buffalo floated here and there to tell the tale. Then, as the waters began to recede, the ghastly scene began to unfold. Estimates of death by drowning grew ever wilder, more unbelievable, ranging from thirty to two hundred thousand souls. Death from disease and starvation would follow. Man's interference with global weather had taken its heaviest toll so far.

We now know that protecting the balance of nature is one of the great challenges of the late twentieth century, and that stable prosperity can be achieved only if the environment is nurtured and sustained. But just how this is to be done remains obscure. Governments can get together to reduce their military might or place a ban on nuclear war, but how do you enforce legislation in the interests of the environment while such legislation appears to be in conflict with the pursuit of wealth? How do you enforce legislation that threatens the short-term profit of companies or endangers the immediate livelihood of individuals? How do you get past the enormous vested interests involved?

There are today an encouraging number of far-seeing industrialists, agriculturists and ecologists who are trying hard to find a balance between these apparently irreconcilable interests, but where a serious clash between the environment and profit occurs, profit invariably wins. An horrendous example of this, presently in the forefront of environmental concern, is the suicidal destruction of the world's rain forests, most of which occur in the impoverished regions of South America and the far East, and all of which are vital to the maintenance of a stable global climate.

If poverty and the environment are inextricably mixed, then finding an answer to either problem means finding an answer to both. So in the process of transition to a moneyless economy it seems to me that there is only one way. *It has first to become more profitable to do things in nature's way than the way we are going now.* In agriculture as in industry, waste must become a prime resource.

The maintenance of soil fertility is the first condition of any permanent system of agriculture. What, then, is nature's way of maintaining soil fertility?

To begin with, mixed farming is the rule, in the forest, the prairie and the seas. Many species of plants and of animals always co-exist, creating niches for themselves in a complex web of living patterns that all interact to enhance the fertility of the soil and the health of the whole ecosystem. Monoculture is unknown in nature. Exploiting vast areas of countryside for the cultivation of a single crop such as wheat or maize has thus been a major departure from natural processes. We'll be coming back to some of the consequences of this disastrous mistake.

Secondly, in natural ecosystems, the soil is always protected from the direct action of sun, rain, and wind. In the forest the whole energy of sunlight is made use of by the foliage of the forest canopy and of the undergrowth, while the leaves break up rainfall into a fine spray for absorption by the litter of plant and animal remains on the forest floor. Moisture not absorbed by the porous soil of the forest floor or the abundant protective layer of humus on the surface drains slowly into the subsoil by means of a network of drainage channels made by earthworms and other burrowing animals. Hardly any erosion occurs. The streams and rivers in forest areas are always perennial because of the vast quantity of water gently connecting rainstorms, subsoil, rivers and the sea in a never-ending cycle of life-giving moisture. 'There is therefore little or no drought in forest areas because so much of the rainfall is retained exactly where it is needed.'[21] In the light of this natural economy the wholesale destruction of forest cited above can only make one shudder.

The forest manures itself. It makes its own humus and supplies itself with minerals. The gradual accumulation of mixed vegetable and animal residues is quietly converted into humus by fungi and bacteria on the forest floor. Howard observes that the process is entirely sanitary. 'There is no nuisance of any kind - no smell, no flies, no dustbins, no incinerators, no artificial sewage system, no water-borne diseases, no town councils, and no rates.' On the contrary, the forest affords 'sufficient shade and an abundance of pure fresh air.'[22]

The mineral matter needed by the trees and the undergrowth is obtained from the subsoil. The dead roots provide the deeper layers of soil with organic matter and an almost perfect drainage and aerating system.The living roots comb the upper 20 feet of soil for such minerals as phosphates and potash which are used in the green leaves. Conversion of these leaves in the cycle of the seasons into humus helps to enrich the surface soil, ensuring that even in regions where the surface soils are deficient in minerals, the soil is exceedingly fertile, supporting abundant vegetation without any addition of mineral manures. Howard remarks that the trees and shrubs of the hedgerows, parks, and woodlands of countries like Great Britain should continue to be used for the maintenance of soil fertility.

> In the future, when agriculture comes into its own and when it is no longer regarded solely as an industry, it may be desirable to embark on long term rotations in which woods and park-land are turned into arable [land] and worn-out arable back into woodland or into mixed grass and trees. In this way the root system of the tree can be used to restore fertility.[23]

Virgin soil always carries a large fertility reserve, the extent of which can be seen when trees are cut down and the land used for agriculture. Howard states that when plants like tea, coffee, rubber and bananas are grown on recently cleared land, good crops can be raised without manure for ten years or more.

The history of the great virgin grasslands of the world reveals the same picture. The grass carpet conserves the rainfall very much as the forest does. Before man moved in there was little or no soil erosion on the grass-covered plains. Vast herds of bison grazed the mixed herbage of the North American prairies. We know that when the early settlers brought the land into cultivation the store of fertility provided by nature was so great that prairie soils yielded heavy crops for many years without livestock and without manure. Under such conditions of fertility, crops and livestock look after themselves. 'Nature has never found it necessary to design the equivalent of the spraying machine and the poison spray for the control of insect and fungous pests.'[24] To our shame be it said that these things are the exclusive concommitant of our own 'advanced' methods of monoculture, whose machines and fertilisers and poison sprays do nothing to keep the land in good heart. 'By their use the processes of growth can never be balanced by the processes of decay.'[25] Howard observes that all they can accomplish is the transfer of the soil's capital to current account.

That this is so will be much clearer when the attempts now being made to farm without any animals at all march to their inevitable failure.

Today, a mere forty years after these words were written we are seeing the results of our mistaken agricultural policies in widespread erosion of topsoil, desertification, lands which refuse to yield any crops at all without the addition of ever greater quantities of chemical fertilisers, disease-prone crops and animals, deterioration in the health of the human beings who consume them, and famine on a scale such as the world has not seen before. In addition the 'waste' products of these ill advised agricultural practices run off into rivers and lakes, seep into underground water supplies, cause untold ecological damage and finally land up beyond our reach in the sea. The contrast with nature's careful conservation and re-utilisation of resources passes belief.

Thanks to the tireless efforts of worried environmentalists, public outcry has already triggered a movement towards sounder practice in industry, as briefly discussed above. But perhaps less widely understood by the general public is the appalling devastation caused to the environment by the misguided agricultural practice of the twentieth century. Monoculture, as opposed to mixed farming, the removal of trees, failure to rotate crops, erosion due to a drop in the humus content of the soil, artificial fertilizers in place of organic 'muck', poison sprays to control insect and bacterial pests, and a host of interrelated practices contrary to nature's ways, have turned vast areas of once fertile land to desert, rendering crops less resistant to disease, farms less resistant to drought, and famine endemic to millions upon millions of people.

Nor is health sufficiently closely tied in the minds of most people to the way in which their food is grown. Were it more widely understood that health in crops, animals and people depends to a large extent on the return of organic wastes to the soil - and there has long been evidence that this is so - public outcry would have been no less intense, and we should have been forced to other than 'economically viable'methods in providing our food.

'We are so made as to derive our sustenance from what was alive so recently that, somatically, it is still living' writes Dr Picton.[26]

Not only is it desirable for our own food, whether animal or vegetable, to be as fresh as possible, but those foods should themselves be nurtured on a living diet. This means that crops should be grown on soil rich in humus to stimulate the activities of nature's free labour force - the millions of earthworms, insects, micro-organisms and fungi on whom the health of a living soil depend. 'History suggests that a decline in soil fertility is always accompanied by a corresponding decline in the vigour of the people who dwell upon it.'[27]

In this regard it is important to distinguish between increase in production and increase in fertility. The enormous increase in production of food brought about by the use of chemical fertilizers seems to have blinded us to this crucial distinction. It brought about the large-scale single-crop farming practices on which the world's food supply so largely depends today - practices which could be quantified and therefore assessed as profitable or not profitable. In the process it cashed in on the existing fertility of humus-rich soil, with disastrous long-term effects, not least among them a loss of resistance to disease.

Every farmer knows that the appearance, in a crop, of parasites and disease is often the first symptom of a loss of fertility, and Howard goes so far as to say that disease is nature's way of removing the weakling crop.

> The policy of protecting crops from pests by means of sprays, powders, and so forth is unscientific and unsound, for even when successful, such procedure merely preserves the unfit and obscures the real problem - how to grow healthy crops.[28]

The need for poison-sprays seldom arises in areas where the tradition of peasant farming has never been broken, as, for instance, among the Mongolian peasants. Ironically, they are mostly too poor to indulge in the luxury of sophisticated modern agriculture by which we ruin our soil. So the age-old wisdom of people living close to the land has prevailed, and every scrap of waste has been returned to the soil.

Lady Balfour points out that in almost every other type of farming to which the West is prone, soil fertility has been lost more quickly than it could be replaced. 'This is due, not only to the excessive quantity of chemical fertilizers now in use, ... but also to the shortage of available organic matter which modern standards of civilized society have brought about, the three main factors being, the introduction of the petrol engine, the movement of the rural population into large urban areas (which the industrial revolution induced) and, perhaps most serious of all, modern sanitation.'[29]

> When horse traction gave way to the internal combustion engine the land was robbed of millions of tons of horse manure previously produced in town as well as country, and when water-borne sewage was introduced into our cities, the capital of the soil - its fertility - which is removed from it year by year in the form of crops and livestock, no longer found its way back to the land in the form of the waste products of the community...[30]

It has been said that Man is the most extravagant accelerator of waste the world has ever endured.[31] Looking around at the 'throwaway' habits of our consumer society with its mountains of litter and toxic waste, at the reckless exploitation of energy and material resources in our industry, and adding to all this a suicidal failure in agriculture to conserve our most precious resource, I marvel how a society which is geared to the notion of economic viability can have gone so far along so strange a road.

It all depends, I suppose, on how you calculate your profit. In the present order of things, economic viability is assessed as though industry and commerce were divorced from the rest of the world, with values important enough to over-ride all else. If you ignore the cost to future men of trying to correct your mistakes; if you use the air and the rivers as sewers and leave it to future generations to clear them up; if you exhaust the available fossil fuels without developing renewable alternatives; and above all if you destroy the living soil on which the health and survival of mankind depends - if, having done all that, you can show a surplus of income over expenditure, you can with justice be considered to be running a profitable enterprise. Few people today are comfortable with such a scene. But it is still the prevailing pattern over most of the world. Drucker observes wryly that 'the future is in the hands of executives who are already fully occupied with the daily crisis, and for whom the daily crisis is the one absolutely predictable event in their working day.'[32] This is understandable. With attention centred on market advantage, on price strategies, on cost accounting, on labour disputes, on faulty materials and the like, the long-range implications of today's small immediacies can well be overlooked. But we dare no longer follow this road. We have to find a way to profitability that takes these long-range effects into urgent account. Using industrial waste as a prime resource, returning agricultural waste to the land, and the sensible use of human wastes, would seem to be some of the ways to true economic viability.

d.) Economic Viability

Our search for true economic viability requires that, in addition to basing industry and agriculture on ecological principles, we need to take an urgent look at the extraordinary divorce of money-making from the basic needs of a community - the 'symbol economy' briefly referred to above, which distorts the flow of capital and tends to concentrate it in fewer and fewer hands. That the rich get richer and the poor get poorer is not just a wise-crack truism. It is a tragic progression affecting more and more people the world over. The marches, civil disobedience, riots, squatting, looting, crime, and suicide, as well as the overthrow of government after government in quick succession throughout the world, are ample testimony that this state of affairs cannot be allowed to go on. Millions of poverty-stricken human beings are discovering that their sheer numbers give them a voice in society - a voice that the rich can no longer afford to ignore. The war of the 'have-nots' against the 'haves' has begun in deadly earnest, and no amount of manoeuvring within the monetary system will make it go away. We have to come to terms with this, not by lending large sums of money to impoverished peoples, nor even by making such sums conditional on sensible practices, but by learning to live as intelligent components of a living sytem. I believe that the day is rapidly passing when people will be prepared to offer their health and peace of mind in pursuit of monetary wealth. The price in terms of security and stress has already become too high.

To do us justice, many caring people are meeting almost daily in national and international fora searching for answers to these ills. But always within the framework of economic patterns which have outlived their usefulness. These patterns evolved in a world which combined sparse population with apparently infinite plenty - a situation directly reversed today - a world moreover that lacked the essential communication techniques which could set us free. Now that we have these techniques, it is time to look again at the human condition, to re-assess human need in real, not symbolic terms; time to say no to any system that cannot foster the human potential in its entirety within a single viable economic unit.

Such an enterprise will entail a deep study in human motivation, and a whole new look at the concepts of work and employment. We'll come back to these. Sufficient to say here that I do not believe man to be motivated solely or even chiefly by a lust for money, and that unemployment in a world needing the labour of millions of willing hands is the ultimate absurdity. Were there no reason but this for drastic economic review, it would be reason enough.

For the moment I want to look at some of the ways in which the use of waste as a resource in both industry and agriculture might engender the embryo of a new information-based economy. The sequence of steps to transition might go something like this:

First a radical paradigm shift toward ecological awareness. After a sluggish start this is now happening the world over with encouraging speed.

Then the realization by commerce and industry that unless they respond to public demand for a 'green'world they are likely to go out of business.

Next might come the realization that this response, by engaging the dynamics of reciprocal waste usage in industry, might actually increase profits.

Concurrently with this development would come a dramatic increase in small concerns arising from specialized technical knowledge in the field of pollution control and reciprocal waste management.

These developmenst are in turn likely to demand and stimulate research into alternative local sources of energy, reducing the demand for oil or coal-based power, and stimulating the construction of many millions of small local power bases.

Meanwhile increased public awareness would engender demand for a change in *what* we produce, *based on advanced knowledge of what we really want and need,* and on dissatisfaction with ecologically harmful practices outside those needs.

The ensuing perception in both commerce and industry that it is in their own best interests to be seen as pioneers in the field, will further advance ecologically friendly practices.

Since reciprocal industry will clearly be more efficient and less costly to operate if materials are exchanged locally than when they have to be transported over long distances, its practice will encourage the development of close-knit activity amongst local communities, supplemented where necessary by materials and know-how from further afield. This in turn will be made more efficient by the free exchange of available information concerning the nearest or most suitable outside source.

This aspect of industry will be more fully discussed in the section on localism, and will of course impinge directly on political issues such as cross-border trade and exchange, and the operation of international corporations. However it will lead inevitably to economic interchange so closely knit from one region to the next, so mutually interdependent, and so interlocked electronically, that it is no longer apparent who owes what to whom. (Do I pay you for raw material for my industry, or do you pay me for removing your 'waste'? Or shall we just call it a day and bypass the cost of accounting?) Money will have begun to phase itself out.

CHAPTER FORTY ONE

9.3 Social Patterns - Land Usage and Taxes

Summary: An exploration of Land Usage, Farming Methods and Small Business as a means to economic and social stability.

Social Patterns

Land Usage and Taxes

We've seen that as in industry, agriculture has developed along lines of land usage directly in conflict with sound ecological principles. But the enormous discrepancies between rich and poor in land usage complicate the issues even more. If we want to get back to smaller, mixed farms run on organic principles, we meet head-on not only with the interests of those farmers and absentee landlords who hold large tracts of valuable land, but with the vested interests of fertilizer, heavy machinery and pesticide manufacturers as well. Clearly any attempt to break up existing farms into smaller, mixed farming entities will involve us in complex and extremely emotive social and political issues far beyond those of agricultural method.

Having in mind in particular the immediate plight of the very poor, it may be helpful to look at this point at a recent South African study by Stephen Meintjes and Michael Jacques entitled *The Trial of Chaka Dlamini - An Economic Scenario for the New South Africa* (Amagi Books, 1990)

These authors explicitly address the complex issues of land ownership and usage in relation to a free market system in the Third World, and are well worth listening to. Their ideas fit in well with my own conviction that none of us can in fact 'own'a piece of land in the true meaning of the term. Whatever else we may be, we are all units in an immensely complex living organism, and though we may be intelligent units able to manipulate (or destroy) the whole magnificent creature to which we belong, the thought of 'owning'any part of it is manifestly ridiculous. As well suggest that a neuron owns a slab of the human brain. Nor can we take so much as a grain of it with us when we die. At the same time we are intensely territorial animals. We value our privacy, defend to the death our right to our own property, and need beyond most other things to do our own thing in our own place and in our own way. How are we to do this while allowing others the same privilege? How may everyone follow his own bent and ambition without stepping on his neighbour's toes? And how can all of us enjoy self-determination and sufficient space without destroying the Earth?

It would seem that in pursuit of individual wellbeing as well as for planetary health, we need a different method of dividing up the land. Socialist doctrine recognized this need of 'the people' for land, but confused the notions of 'state'and 'people', and followed the calamitous route of state ownership and control. This route is now closed. In the last decade we have seen its almost total collapse, to the accompaniment of untold suffering for millions of people. But the crisis of the poor has only just begun, and our own collapse is not very far behind. Unless we can find a better way of dividing the land, endemic poverty will end in destroying the Earth.

In the context of my own work then, the problem at issue is how to make the change from vast areas of arable land devoted to monoculture (owned by a small land-rich section of the population) to smaller organic-type mixed farms owned by very large numbers of ordinary people; and (as a step in the direction of a moneyless economy) how to enable large numbers of people who have neither land nor capital to set up viable small business interests in the communities where they live. Put in another way: How, in societies with many hungry mouths, where most of the people live in abject poverty, do you create a situation in which anyone can have security of tenure over a piece of land appropriate to his need and competence?

South Africa is a particularly useful proving ground for work in this sphere, because it is at a stage in its history of acute crisis and constitutional upheaval, to the accompaniment of considerable instability. We've seen before that this is a situation not only conducive to change, but perhaps even a necessary prelude to major evolutionary development. The wrong decisions now could precipitate disaster. Creative thinking could usher in a wholly new kind of society, unique to local conditions and geared to the emerging paradigms of planetary health.

Meintjes and Jacques suggest that the answer to land hunger in South Africa lies in replacing taxation by a land-usage rental based on the market value of un-improved land.

Very simply the suggestion is this:

All existing forms of taxation should gradually be phased out. In South

Africa these include at the time of writing, an income tax on companies and individuals and a number of indirect taxes such as GST (General Sales Tax) - soon to be replaced by VAT (Value Added Tax) - and customs and excise duties, all of which fall most heavily on people with drive and initiative, whose capacities and labour enable them to generate economic growth, and on the very poor, to whom any form of tax on their earnings can mean real hardship. Thus taxation is seen as a disincentive to enterprise, an unfair burden on those with a flair for production, and a brake on the potential prosperity of the land. These taxes should therefore gradually be replaced by a single source of communal revenue, based on an annual site-user charge on the unimproved value of land, this amount deemed due to the community (and hence to the government) by virtue of society's having created the infrastructure, opportunities and value which only society can provide.

This user charge would be based on the current market value of the land, a concept first put forward by Henry George and endorsed by others as far removed from each other as Leo Tolstoy, Winston Churchill, Sun Yat-Sen and Milton Friedman. In the hands of Meintjes and Jacques, George's fundamental idea is combined with traditional tribal land tenure, and emerges as an unique, historically based, thoroughly African solution to this most thorny of all social problems. It is a marriage between land use without speculation, security of tenure for the individual and a gateway to fair opportunity for all - a situation with which, prior to reading this book, I had wrestled unsuccessfully. I salute the authors for their insight. They present their ideas through the voice of an old Township African, Chaka Dlamini, on trial by his radical peers for alleged treason to the cause of freedom. This device enables the authors to voice the economic needs and aspirations of a number of widely diverse sections of the community, and in particular those of the underprivileged. That in the case of South Africa a vast majority of the underprivileged are black, in no way detracts from the potential universality of the system, nor from its adaptability to local conditions in other parts of the world.

Old Chaka suggests, then, as follows:

> We must pay, and our government must collect, the natural rent on land. Everyone who has land, whether peasant or panel-beater, mining magnate or manufacturer, shopkeeper or shoemaker, must pay an annual rate according to the value of his or her land. If it is land with gold, perhaps he will have to pay millions; if it is land in Eloff Street, perhaps hundreds of thousands; if his land is in the platteland or the

homelands, perhaps he will pay nothing. Remember, it is the people of South Africa, black and white, who over the centuries have given to every square inch of our country the value that it now has. The man who works well and hard on good land will make more than the man who works equally well on poor land. That 'more' is the people's because it is the natural rent arising out of the difference in the value of the land.[1]

Chaka gives as an example the case of one Mbezi, a very successful Soweto business man who runs a 'Shebeen' (an African 'pub'). Mbezi knows his business; he charges fair prices; his place is clean and well run; the people who work for him work hard but he treats them fairly. That is the part of Mbesi's success which is due to his own character and efforts.

But there is another reason why he is successful: he is in a good part of Soweto. Not the best, but good. People there have good homes, some have bought their houses and most are thinking about doing that. Because of the doctors, lawyers and other professionals from the neighbourhood who have a drink there, it has become known as a high-class tavern. Move Mbezi to some other part of Soweto that is not as prosperous as this part, or to a far-off country village, and with all his ability and hard work he would not be as successful. So some of his success is due to his own efforts and some is due to where he is. And that is the part the community should look to for its revenue.[2]

In the given context the argument is irrefutable. However I have reservations concerning the general supposition that the value of land is wholly (or even chiefly) man-made. In terms of human survival or the ecological health of the planet, an area such, for instance, as the Amazonian rainforests, is clearly beyond all price, and should be protected and nurtured by all humanity, whatever the monetary cost needed to afford economic opportunity to the peoples in whose territory it falls. However the value of the rain forests and other crucial wilderness areas lies in properties quite other than man-made economics, and in the interests of human well-being their preservation requires a very special approach not at issue in the study by Meintjes and Jacques. The special value of untouched ecosystems does not therefore detract,I think, from the validity of the argument.

How is one to determine, then, the monetary value of land, in such a way that both rich and poor can afford to pay this 'user charge', and how can re-distribution come about so that everyone who is prepared to work his land can have tenure over an acreage sufficient to his needs?

These questions take us to the very heart of capitalist and socialist argument, and in particular to the basic requirements of, and access to, the means of production. Given that the three basic factors of production are land, labour and capital, it is very clear that an overwhelming percentage of human beings the world over simply haven't a hope under any of the systems on which the world operates today. If we are to find our way to a more equitable system of land tenure and small business development, while at the same time moving towards sustainable methods in both industry and agriculture, we have to involve the capacities and self-interest of every man, woman and child on Earth, and this, as it seems to me, means giving very large numbers indeed the privilege of using a piece of land.

Meintjes and Jacques quote a Sotho saying that whatever a man is, he is on account of other people, and argue that this is how it is with the land; 'its value is on account of other people, and [a man] has the right to that land because other people allow him that right.'[3]

The orderly conduct of any democratic city or country depends on an agreed system of law which endorses a man's peaceful authority over his own property, and his right to pursue his legitimate business without interfering in, or interference from, other people. Not only do people agree not to take away each other's land or its produce, but 'his land enjoys many things other people have done over the years, such as build roads, bridges, railways, airports or large cities'. Even if his land were bare it would benefit from these things, and would be corrrespondingly more valuable, which means that in this sense the unimproved value of the land itself rests squarely on the efforts of other people.

Chaka sites three duties inherent in a man's peaceful enjoyment of these rights.

> He must keep his land in good condition, must not disturb his neighbour's enjoyment of their land, and he must pay the rent.[4]

These rights and duties of the individual are balanced by the duty of authority (government):

> To keep unoccupied land in good condition, make unoccupied land available to anyone who wishes to make use of it, and finally, to collect the rent.

Once again I have reservations about making unoccupied land available to anyone who wishes to use it. In the ecological context there are sensitive areas of such inestimable value to the community as a whole, to the richness and diversity of the ecosystem, and even to the healthy functioning of the whole Earth organism, that the wish of a single individual cannot be allowed to take precedence. This aspect of decision-making will be discussed in some detail a little later on. Sufficient for the moment to notice it in passing.

Getting back to the question of the allocation of land, and the proposal to charge a user rental in proportion to its unimproved 'value', how does the community decide whether one piece of land is better than another? Who makes the decisions on behalf of the community?

And how does one prevent endless litigation on the justice of the values set?

The answer, in a market-oriented society, proves surprisingly simple:

> 'the community' is determining relative values all the time by means of market prices paid for land. An acre of land in the middle of Johannesburg will sell for more than R40 million, while in the far north-western Cape, you would be lucky to get R40 an acre. So the natural rent or usage charge for land in the middle of Johannesburg would be far greater than the charge for land in the Kalahari, because, as the market values tell us, the one may be as much as a million times more productive than the other.[5]

The central thesis of Chaka's doctrine, is that tax on work of any form, 'whether labour, management, enterprise or research, is not merely lacking in common sense, it is counter-productive and unjust.'[6]

In the period prior to Zimbabwian independence, Lord Soames, the last governor of Rhodesia, is reputed to have said that 'if one were to set out with a specific, stated objective of designing a tax system which would act as the maximum disincentive to effort at all levels of society, which would penalise and deter thrift, energy and success, it would be almost impossible to do better than the one we have in this country today.'[7] Zimbabwe, nearly two decades later is still wrestling with these problems, and is, at the time of writing, embroiled in problems of land ownership very similar to those at issue in South Africa. Nor are they endemic only to Africa. The economic problems of the so-called 'Third World' anywhere on the globe are very similar to our own and will demand creative thinking of the highest possible order for their solution. They will also demand a universal 'will' which is at last beginning to emerge in economic circles.

The issue is how to ensure equitable conditions for the poor of any land without discouraging those whose past energy and drive have not only entrenched their own position, but have contributed significantly to the creation of wealth. Once again we have a head-on collision between the 'haves and the 'have-nots. Thanks to the ubiquity of electronics, poverty has become so highly visible to everyone that not the most sanguine self-interest can make us fail to see its eventual impact on ourselves.

Perhaps it is necessary to remember at this stage that even a fairly equitable monetary system of the kind envisaged by these authors is not the final goal of my own thinking. The reader will remember that we are moving slowly but surely towards the vision of a moneyless economy in which economic self-regulation follows organic patterns common to all living systems. The suggestions presently under discussion must therefore be read only as possible means to transition, first steps from any current arrangement anywhere in the world to the synergetic cohesion of a living entity. There should, and will, be many such. But even were the need for money miraculously removed from the world's poor, people with property would still have a head start on people with none, a situation on which I had spent considerable unsuccessful thought before encountering the work of Meintjes and Jacques. How, under any economic system, does one achieve equality of opportunity? And having achieved this idyll, how can it be maintained?

I have no illusions about equality as such, either of character or of ability, but we are looking for a situation in which everyone has the opportunity to maximize his own potential. This means that whatever his interest or bent may be, everyone who wants it should be enabled to acquire a piece of land suited to his needs. For many rural people, this might mean no more than that a man has a piece of agricultural land large enough to grow food for his family, with something over for the purchase of goods he could not produce himself. For others, like Mbezi, it might mean an appropriate site on which to run his business, and a piece of ground on which to build a home. The variations in terms of personal need or desire are infinite, but most of them involve secure tenure over a piece of land.

In the Chaka Dlamini scheme, rural people living on land with a low market value would pay little or no user rent, and could therefore use their produce to the full, feeding their own families and selling the surplus. The user of valuable business sites in a big city, on the other hand, would pay more because he was enabled to cash in on the activities and initiative of his fellows.

But how would anyone presently owning nothing obtain such land? To simplify a rather complex question, there are two essential steps to this.

Firstly, all privately owned land would be put up for auction as soon as the owner wanted to sell.

Secondly, banks would be authorized to lend money, without security, to trusted people deemed capable and determined enough to prove worthy borrowers. In other words, the criteria for a loan would shift from 'security'to production. Such an idea is based, like so many of those we have been discussing, on localism - small groups of people who know each other and therefore know amongst themselves who is trustworthy, who is hard-working, who is slovenly or unreliable - in short, who is worthy to receive a loan. At a single stroke the basic disadvantage under which the poor presently labour would have been removed. No longer would the banks be lending only to the rich - the people who had security, but to any man, rich or poor, who enhanced useful production.

Meintjes and Jacques quote an interesting programme on TV about the Poor Bank in Bangladesh.

> It was started by a Bangladesh man who went back to his country with an MBA and a PhD from an American university, and who had worked in banks in the USA. But when he arrived home he threw away his textbooks because he saw they were of no use in meeting the needs of his people, many of whom are even poorer than ours. Instead, he started this Poor Bank which lends only to the poor, and only for production.
>
> All the borrowers are members of a group of friends who meet once a month with the banker who looks after their group, to discuss their needs. One woman, with seven children, and with an annual income of $84, was lent money for a loom. When this TV programme was made three years later, she had fifteen people working for her, and was earning $3000 a year.
>
> Because the borrowers meet regularly with their group and the banker, they trust one another. As a result, they only take money for capital items they can use properly, and there are very few bad debts.[8]

Once again there will be many variations on this theme, some of which are already under way in Africa. Their significance lies in the principle of localism, found throughout my own work to be crucial to self-regulation along organic lines - in oblique barter, reciprocal industry, small mixed farming operations, appropriate technology, renewable energy, face to face discussion, exchange of information - and now banking. It only requires the effective diffusion of power to local communities, and we shall be well on the way to establishing a groundwork for the self-regulation of society as a whole. We'll be coming to this.

Meintjes and Jacques suggest that because the user-charge would be exorbitantly high on valuable land, it would no longer be to the advantage of the speculator to hold large tracts out of circulation, and much useful land would come on the market. Its value (to him) would therefore come down, while retaining its usefulness to the community at large, and many people would be eager and willing to acquire tenure of a piece of it and pay the appropriate site charges. On the other hand less-centralised or otherwise 'less-valuable' land would command a low market value and carry a correspondingly low site-user charge, thus enabling the very poor to acquire a site, as well as to pay the user-charge on a continuing basis. Some might even be able to acquire marginal land without any 'natural-rent' at all.

Let us suppose, then that a man is able to acquire a loan to buy himself a small farm. And let us suppose that a suitable piece of land has come on the market. The land is put up for auction. The market determines the price, and the buyer, having convinced his bank that he is a good risk, acquires tenure over the farm for as long as he wants it - perhaps for life - on condition that he both keeps it in good condition and pays the natural rent. Gradually, as more and more land comes on the market and more and more poor people are enabled to acquire it, the present imbalances begin to even out - and this without attempted coups or the violent seizure of power so prevalent today. In addition, starting from where we are now, large tracts of suitable farmland, presently owned by government, could be brought into use.

CHAPTER FORTY TWO

9.4 Organic Principles in Politics

Summary: As in living organisms, we should each and all be integral working parts of the body politic.

Representative government is found to be an inadequate expression of this dynamic.

Alternative methods are explored.

Organic Principles In Politics

In every sphere touched upon so far in *The Cosmic Microbe*, we seem to be moving away from power blocks, and towards the role of the individual in the greater living organism. This is important, because it involves the concept of individual responsibility. As long as one can 'pass the buck' to some greater authority, and ultimately to the government, the less duty there would appear to be on the individual to accept that he or she is absolutely responsible for his own actions, and hence for the quality of the system to which he belongs. The systemic approach belies the notion of hierarchical authority. It insists that society is a living system composed of human beings and social organizations in continual interaction with one another and with the surrounding ecosystems on which our lives depend. Our social and economic systems are thus not so much structures as relationships in motion, and these relationships include our interaction with whole galaxies of interweaving environmental networks.

We've seen that in agriculture as in industry the nonlinear interconnectedness of living systems imposes checks and balances on the optimal size of any structure, organization, and institution, whether farm, factory or business monopoly. We've seen that maximizing any single variable such as profit, efficiency or GNP for example, to the neglect of ecological principles, will eventually, as with the polyploid horse, become unworkable. We've seen how industrialists are beginning to apply the systemic approach by embracing ecological principles. But the thorny question facing each and every one of us remains:

How to contribute towards the organization of a society in which each individual has freedom to fulfill his own potential without making any intrusion on the freedom of his neighbour. And this brings us to the vexed question of politics, which I had most fervently hoped to avoid.

As I see it, political science is an attempt to impose structures on human relationships which will enable society to pursue an orderly and functional course, in spite of the determinedly self-centred nature of each individual member. The so-called 'lesser' living organisms on planet Earth seem to have found a workable and beautiful way to achieve this miracle - a miracle we are only beginning to explore. The key would seem to lie in realization that everything anyone does or thinks involves us. Since we too are working parts of the system at every level, our lives depend on workable and concerned relationships with everyone and everything. So important, indeed, is this relationship that we can no longer afford to delegate our authority to others, and the whole question of representative government comes in question.

So far as I am able to understand, living processes do not embody any form of representative government. Is it possible, therefore, to find a way in which human beings, too, for all their individualism, can operate as functional parts in a living process?

Once again, I believe, the key lies in local interaction between individuals, embodying the twin principles of diffusion and communication that run unbroken through all the discussions in this book. In nature this implies the slow, interlocking processes of all the planet's life-forms; in the universe, the universal ripple of interacting energies. In society it would seem to include, among other important issues unique to the human species, the manner in which society regulates its behaviour patterns and norms. In most societies this process has culminated in that rather arbitrary grouping of millions of human beings we know as the Nation State, with its top-heavy hierarchical structures and representative forms of government.

In any of the basic activities of mankind on which we have touched so far - such, for instance, as agriculture, commerce or industry - it is clear that our present manner of operation no longer works (if it ever did) for the vast majority of our citizens, and that the repercussions of outmoded methods are having global consequences too dangerous to ignore. It is not illogical, therefore, to ask ourselves whether perhaps political systems might not be in equally serious need of overhaul.

We've spoken of human beings as units in an emerging global brain, and observed how some brain processes follow essentially 'bottom-up' principles which 5th Generation computers are beginning to emulate. While early computers were linear in character, basing their efficiency on their speed of operation, computers now use parallel processing, pattern recognition and related technologies that in part emulate the complex and distinctly non-linear processes of living tissue. In just such a way, I believe, must we reorientate our thinking to evolve systemic rather than linear thinking in our politics. Hierarchical structures must give way to a diffusion of power in bottom-up local units, which in turn embody the synergetic interaction of individuals.

Is there any way in which we can do this?

The first step would seem to be the diffusion of central power to small local communities, based on groups of people who know each other and therefore most naturally interact. These in turn would interact with larger groups who would interact with ... all the way up the line, so that the outcome of all the social adjustments along the way would embody each and every individual in its final outcome.

While this would seem to be the way in which single living cells must have come together to form organisms, it is the direct opposite of the mode of government of the Nation State. National governments mostly consist of elected members representing large bodies of people. They exercise authority from the top down; and their followers have little access thereafter to decisions directly affecting their lives. Protest marches, civil disobedience, violence and mass action by increasingly large numbers of determined and sometimes angry people all over the world, signal their dissatisfaction with this state of affairs. In spite of attempts at unification such as the EEC, or of world bodies falling under the United Nations authority, more and more fragmentation of nations is in fact taking place, signifying the desire of smaller groups to pursue their own affairs in their own way, and of individuals within these groups to have a say in the conduct of their lives. The breakaway of the states formerly in the Soviet Union is a case in point.

If society is to be self-governed and the Nation State to lose its position at centre-stage, we are faced with enormous numbers of complex and interrelated questions. What kind of structures do we still need? What parts of the old structures will still have value? What can we dispense with with impunity - or at least to society's ultimate benefit? Do there still have to be laws governing our conduct towards each other and society? Shall we still need the courts? How will we regulate our schools, and who will decide what shall be taught in them? Shall we still need schools in their present form? And what about things like health services, hospitals, old age and infirmity? Or mental health? Who will care for the sick, the disabled, the mentally ill? And at whose instigation?

All these things and many others need serious creative thinking, not in the sense of problems to be solved, but in full realization that all the parameters have changed - indeed *are* changing exponentially day by day to the point that nothing short of a dynamic new life-type approach will do.

Self-government even at local level will have to be cybernetic in principle, acting on feedback. Cybernetics is the theory of communication and control in living beings as well as in machines, so I think our best guide will always be: 'How does a living body do this thing?'

We have been schooled for so long in political thinking, and are so used to the idea of the Nation State, that it's hard to break out. One thinks of the Nation State as immutable, as having existed always, which it has not. And when trying to imagine alternatives one thinks automatically in terms of 'replacement' of the existing system. It doesn't have to be replaced. It must and will evolve, the way a group of cells must have evolved toward hepatic, glandular or neurological function. Already its authority is being substantially overridden in the economic sphere by vast international corporations, whose ramifications are beyond the power of any single nation to control, and while it is doubtful whether vast monopolies in either power or wealth can survive the new trends, their capacity to sidestep national authority is clear. They also largely control government policy through financial pressures. Again a number of questions spring to mind:

- In how far can giant international economic units evolve to the point where they effectively handle, on a global basis, the essential functions now handled by governments, rendering superfluous the Nation State?

- In how far is it desirable for them to do so?

- Have they a useful function in the long-term interest of the planet?

- How, if at all, would they complement or otherwise interact with a self-governing society based on systemic principles?

- In the areas in which these big concerns already work, such, for instance as energy, armaments, banking, mining, shipping, aviation, steel manufacture, car manufacture, railroads, timber, hotels and tourism, and last but not least, computers and telecommunications, to what extent are their interests sustainable and desirable in a society based on living principles?

While I foresee considerable modification in many of the above areas of international financial interest, it is significant that computers and telecommunications, increasingly the life-blood of international commerce, may be the very means to resolving the whole appalling mess in which we find ourselves.

We have to ask ourselves, then, how these international 'empire-building' wizards of the financial world can re-direct their energies towards organic-type activity in the interests of the planet as a whole? And how can they best contribute to a new political paradigm?

Most of these questions are way beyond my competence to answer, though I think they will resolve themselves naturally in the process of bottom-up re-orientation of society. The point I am trying to make is that the power-based structures of the Nation State are already being substantially undermined, that the present state of affairs is dangerously inoperable, and that the bottom-up alternative under discussion in this book is therefore not as bizarre as it would at first appear.

In politics, then, as in agriculture and industry, is local action within a global ecological framework a practical alternative?

CHAPTER FORTY THREE

9.5 Political Diffusion in the South African Context

Summary: The troubled matrix of South African Politics yields some interesting ideas on methods of political diffusion.

Diffusion In The South African Context

There is an old saying that necessity breeds invention, so not surprisingly, the troubled matrix of South African politics in the eighties and nineties once again yields some interesting ideas.

In his book *Permanent Peace* (Saga Press, Johannesburg, 1985) Denis Beckett sets out his aims in the following words:

> The target I am seeking is one in which the vast majority of people of all complexions conduct their lives the way they choose with contentment and faith that they are getting a reasonable deal from the political structure.[1]

This is not too far from my own notions of political adequacy, with one important difference. It is time we moved away from what we are *getting* to what we are *giving* - away from 'rights'and toward contribution. It is time, and more than time, that we stopped demanding from authority what no government is presently able to supply, and took responsibility for our own lives. So I would paraphrase Beckett's words to my own purposes something like this:

> The target I am seeking is one in which the vast majority of people of all complexions conduct their lives the way they choose with contentment and faith that they are contributing to the overall good.

In today's political climate this entails a major paradigm shift, but a necessary one if we are not to destroy ourselves. The Gulf War, so lately over, is evidence enough that in spite of widespread understanding of impending ecological disaster, we still put national, corporate and personal interests ahead of those of the Earth on which we live. We do so at our peril. But perhaps the very horror and wastage of this latest conflict will finally turn our minds to a different approach, particularly in the sphere of energy. We'll come back to this.

In the political sphere Denis Beckett is speaking specifically of and for South Africa, with existing political parties as his starting point.

> When the democratic process begins, (from local level) it does not start from a tabula erasa. *It starts from the basis of what is*, which is that the 'haves' have almost all the land and the wealth and the have-nots practically none. In one way or another nearly all the political players wish to amend this situation, with various different motives in mind ...[2]

This could equally well be said of the poor of any country in the world, and is central to my own striving for economic reform, but I hasten to say that I am ignorant of this author's thoughts on wider political issues. Nor does he give any indication in this work of interest in ecological concerns. But his insistence on the need for a bottom-up structure in South African politics fits well with my own global interests, and provides a useful groundwork in any society for transition to a self-governing synergetic community.

The first step, says Beckett, 'is to provide for the inauguration of local councils. These councils are thoroughly local. Within a large city there may be several councils, based on groups of neighbouring suburbs or even suburbs themselves. In a rural area a small town and its surrounding farmland may be a unit.'[3]

> These councils have considerable powers. In fact they have the power to do anything they want. Unlike the existing local authorities, whose basic function is to act as administrators of dog licences and housing permits within the framework of central government policy, they can make decisions which materially affect the lives of the people within them. They can decide for instance who they wish to admit to their membership and on what terms. They can decide what kinds of industry they will allow to operate where. They can decide on their own tax base, on the degree of conformity or liberty they permit within them, on whether their nature is socialist or capitalist. *They can decide on anything at all.*[4] (The italics are mine.)

The reader is immediately struck by obvious practical difficulties in providing services on so acutely local a basis, and in the financial problems such a system must engender. In particular the intricacies of a multiple tax base would seem to present formidable difficulties. However, let's look a little further at what Mr Beckett has to say.

> Naturally, this is not a very egalitarian, or even fair, dispensation. Each council's freedom of action is limited by what it can pay for, and there is a discrepancy between the benefits which a power-wielding

> local council provides to the residents of Sandhurst, where the average household income is something over R100 000 a year, and those which the same facility offers to the residents of any of innumerable homeland resettlement areas ... where the equivalent figure is not far removed from R100.[5]

In the context of my own work (which embodies the final goal of a moneyless society), this limitation would eventually be phased out, but in the transitionary stage it would clearly give grounds for discontent. Beckett goes on to say that while many local councils would be ethnically homogenous, there are many areas where people of very different backgrounds work together, notably the commercial and industrial districts of the cities. He suggests that these areas too should qualify as local authorities, and that, in effect, this would mean each person working in them would now have two votes, one in the area where he lived and one where he worked.

Step two in this scenario would be the creation of regional authorities consisting of perhaps twenty local areas. On precisely the same principle as the local councils, regional councils would be empowered to exercise control over anything they wished, *to whatever extent they were able*. The italics are mine, for here, early on in the process, occur the constraints and accomodations with which we are familiar in natural processes. Beckett observes that 'Division of tax resources as between regional and local authorities becomes a matter of fluid interaction between these authorities in each case, very likely resulting in substantial differences between one region and the next.'[6]

The mind boggles, but no matter.

Division of powers would follow the same lines, resulting in considerable variation in the degree of effective control, as well, I imagine, as in the measure of content. However Beckett observes that in all cases two imperatives apply:

> *Firstly there are always some areas of activity which the local grouping cannot administer in isolation, and for which it has to have some degree of cooperation with the broader formation around it.* Secondly, no matter how self-contained an individual local area may be, it is compelled to act always with an eye to the possibility that if its behaviour should become sufficiently offensive to the majority of residents of its region, its self-containment is threatened.[7]

Step three is the repetition of the same process on a yet broader scale. Beckett suggests that groups of contiguous autonomous regions form into supra-regional authorities, and so on, as many as prove necessary in each case, and that in each subsequent grouping a man should be entitled to a further vote in the the conventional one-man-one-vote fashion of present-day democracy. The voting power of each adult would thus be felt throughout the developing structure, right through to national level. I find myself wondering whether such a vote, in a community become aware of ecological priorities, might not indeed be a potent force for the common good. However I would like to see some system of direct individual participation rather than representational government.

Beckett makes the point that the determining factor of political life would become one of what 'in practice' people perceived to be of benefit to themselves. Whatever attitudes, similarities and differences there may exist among people and parties, 'they can best be contained by a structure which combines autonomy with interdependence.'[8] This is entirely in line with my own perception that self-interest is a healthy biological principle, and that the proper integration of individual will and capacity is the soundest basis for a stable society. In the 'intensive democracy' described by Beckett there is no 'ruler ... There are instead large numbers of partial rulers and all of them are continually interacting with other partial rulers who have powers and a power-base of their own.'[9]

It might be better to arrange for individual voting on pertinent issues at regional and supra-regional levels, rather than by representation as Beckett suggests. The technological capability of electronic networks is already so far advanced that no complexity of political interlocking should prove difficult. Either way the essence of Beckett's suggestion is sound. However small the primary local authority - and 'perhaps there is no such thing as too small'[10] - the population as a whole is bound together in a fabric of interlocking interest and pressure groups, 'all of which are constantly interacting upon one another according to an endless range of imperatives'.

Now this kind of political structure clearly leaves room for constant fluctuation, and because of the multiplicity of its interacting forces, has a quality of 'flow' as opposed to 'structure'. In this respect it resembles closely the systemic interaction of energies we find in nature, and is therefore one step closer to the bottom-up self-regulation I have in mind than are the rather clumsy hierarchies of an elected Western style democracy.

Beckett argues that what he calls 'intensive democracy', as described above, would do much to even out the enormous gap between rich and poor through a process of political accomodation. His approach to the vexed question of land distribution in particular, rests on political structures devised to encourage maximum interaction between people and groups, and maximum interlocking of conflicting interests. In contrast to this approach, Meintjes and Jacques suggest, in *The Trial of Chaka Dlamini*, the 'user-rent on land' discussed above.

It seems to me that a combination of the two systems deserves serious thought. The elimination of all forms of taxation other than a site-user charge (suggested by Meintjes and Jacques) might greatly reduce the administrative confusion of a multiple authority as put forward by Beckett. Moreover Beckett's suggested political structures would seem to have a better chance of success stripped of financial constraints.

Of course the collection of taxes and their allocation are two quite different things, and while Chaka's suggestions concerning a 'fair' method of taxation may have considerable merit, it does not overcome the thorny problem of how such public moneys are to be allocated. Nor does the continual and shifting debate amongst interlocking local authorities put forward by Beckett on this all-important matter suggest a peaceful or constructive solution when it comes to the apportioning of financial resources. It is difficult to see how effective services could be maintained if the allocation of moneys were subject to ongoing discussion between small groups of people, each with his own particular interests at stake. The way I see things at present there may indeed not be a simple solution to this conundrum *within the monetary system.* Both these authors, like myself, are seeking a means to end poverty and inequality of opportunity, Meintjes and Jacques through a market-oriented 'Land-tax' system, Beckett through diffusion of political power in multiple interlocking political structures. When we move to the information-based organic society which is the subject of my own study, most of these rather daunting problems will disappear, but we have to find our way through the transition period. Robert Heller suggests that 'the way out of the information maze is to establish where you want to go.'[11] and this essential bit of advice would seem especially applicable to the vexed matters here under discussion.

Allocation Of Public Revenue

In the transition, then, to a moneyless, information based society, where do we want to go?

The complexity of the situation engenders a multiplicity of options, which in turn put pressure on decision-makers at every level, but while individuals and small groups know well enough what their needs are, they do not always have access to adequate information on which to base their decisions. Decisions are therefore taken daily on what Heller calls 'inadequate information inadequately analysed.'[12] While technology has the undoubted capacity to cut through this dilemma by supplying relevant information at the touch of a button, during the transition period in which we find ourselves in the nineties, and especially in the Third World, we have little access to such refinements. We have to find a way so simple as to resemble the universal laws on which nature herself operates, and so personal as to engage directly the talents and potential of vast numbers of people. Not even the most benign of democratic governments is geared for this. We all now that in business as well as in politics, individual performance is never better than when designed and motivated by those who are engaged in it. The removal of this immediacy by delegation of authority is one of the great weaknesses of democracy, for while each man may have the right to vote, his power to direct the course of events thereafter is severely circumscribed.

What we are looking for, then, in the transitional period, is a method of allocating state funds, as simple in concept as the 'user-charge'on land discussed above - and so direct that each man would be enabled to have a say in how his share of such moneys should be spent, subject only to the restraint that his decision be in line with the best ecological interest of the planet as a whole. Just how this last condition might be satisfied will be the subject of a separate discussion. Let us look for the moment at one possible means of handling the national budget that would involve Mr Everyman in responsible decision-making, and help to bridge the gap between people's desires and expectations and the actual resources of the country:

Supposing that the diffusion of power to local authorities along the lines suggested by Beckett were combined with the 'natural-rent'system of taxation suggested by Meintjes and Jacques, and to this were added *the allocation of government funds by the counting of heads, including children, in each area of local authority*? With the exception of funds for the kind of service only government can supply, would this kind of proportional allocation enable local authorities to satisfy the reasonable demands of reasonable men?

Such an arrangement would without doubt serve to focus responsibility on a wide spectrum of people, and to force on them an awareness as to what could and could not be done within the limitation of resources available to them. It would also involve every adult in decisions concerning his own priorities, while making it abundantly clear to all and sundry that the creation of wealth over and above such allocation was very much up to him. It would modify the prevailing myth - natural enough in people who own almost nothing in the midst of plenty - that wealth is a stock which can be 're-distributed'. It would tend, by involvement, to promote the notion of wealth as 'a continual process of creation', as much dependent on every citizen as on every other. And this understanding, by releasing the atrophied power of the individual, might prove the most useful tool of them all.

Take the provision of schools, for instance. The available moneys would automatically be proportional to the number of people, including children, in any local area, and it would be up to the local councils, in close cooperation with local people, to decide how much of their funding was to be spent on education. In a young community, with many children, schooling would then receive as big a proportion of the local (and inter alia, national) budget as the community could afford.

If, on the other hand, the local community consisted predominantly of older people and there were few young families, fewer schools would be required and perhaps more attention would be paid to the needs of senior citizens. The significant factor in all this would be that local people know local needs far better than central government can possibly do, and are thus in a better position to allocate their share of the national cake efficiently.

If the area boasted hot springs, or abundant sunshine, or wind or wave power, and thus had easy access to alternative forms of energy, less of its available moneys would need to be allocated to wider networks of electrical power.

If the community were relatively self-sufficient, say a prosperous little community of small organic-type farms, and only required moderate access to road or rail links, it would need to contribute relatively little of its funds to the building of national roads.

If on the other hand it boasted a thriving tourist attraction, and its prosperity depended on adequate transport facilities, it would probably be happy to contribute substantially to access facilities either by rail, road or air, or all three.

Large cities, with a high head-count (who would in any case have paid high land-user-charges to the community) would have need of a high percentage of national expenditure on items such as schools, hospitals, recreational facilities, road networks and the like, and would find themselves, by virtue of the number of people living there, in possession of a large and proportional slice of the national income.

One remembers that, in Beckett's scheme, places of work also qualify as local authorities, so that for instance, in a busy industrial area, 'heads' would qualify for the second time in the count for allocation of national funds. The additional funds needed to supply power lines and other amenities would therefore be proportionately available.

What about the apparently insoluble problem of squatter camps in and around the big cities? One hopes that as land became more readily available, and more attractive life styles opened up, the inexorable drift to the cities would lessen. Meanwhile we are faced with vast numbers of faceless people without employment, without hope and often without food or shelter. Here again the head-count allocation of national resources would seem to be a partial answer, and local squatter-camp councils would presumably opt for housing as a top priority. These communities do seem to pose problems, however, in the Dlamini scheme of land-rent charges, since they occupy high-value sites which benefit from the presence of other people, without being in a position to pay the natural rent. Partial answers which offer some hope, would seem to lie in the site-and-service schemes under way here in South Africa and elsewhere. Peter Wilsher and Rosemary Righter observe in *The Exploding Cities* that:

> Rapid and effective improvements in the material and social conditions of urban life depend on the liberation of the people's own vastly underrated capabilities ... If people who are illiterate and who often live in appalling conditions can make a reasonable and often imaginative job of coping with circumstances, then more rapid and more effective change depends on their freedom to do what they are capable and willing to do for themselves.[13]

Given access to a more favourable division of land after the manner of the Dlamini scheme, and given the authority of their own local councils to use the allotted government funds as seems best to *them*, impoverished people might pull themselves up by their own boot-straps more quickly than we imagine.

The spread of communication in the form of radio and television networks, and the availability of good land at minimal user-charge rentals, might in some cases prompt a return to rural life. Others might feel more at home in the cities. But either way the emphasis would shift from reliance on governmental hand-outs to a sturdy self-reliance, which is the inescapable dynamic of all healthy societies.

What about the other areas on which National States spend public moneys - Defence, Policing, Pensions, Prisons, Administration and the like?

With the diffusion of authority to innumerable interlocking local and regional councils, and the consequent large-scale distribution of public money to these bodies, much of the need for macro-spending by government would fall away. It does seem likely, too, that with the growth of personal responsibility and pride in the community, together with the opening of opportunity and the retreat of poverty, the need for both a large police force and large-scale prisons would be reduced. In addition the national police force might be partially replaced by more appropriate local officers who knew the local people.

As to Defence - if any further evidence were needed, which it is not - that the world's greatest dangers lie in the area of so-called 'Defence', the Gulf War should surely have disillusioned us. The systems designed to defend the sovereignity and interests of Nation States (and especially of oil) now threaten our very survival, while only the concerted effort of all humanity will serve to re-establish the life-support systems of the Earth which we ourselves have destroyed. The time is surely overdue that we get our priorities right.

CHAPTER FORTY FOUR

9.6 STOP AND TAKE STOCK

Summary: Much work is found still to be done to turn these ideas from so much 'Pie-in-the-sky' to a workable reality.

STOP AND TAKE STOCK

How far have we got in our search ?

Do we know at this point where we want to go?

It should be clear by now that for me the search is towards an organic self-regulated society in which poverty disappears; in which information replaces money as the driving force in the economy; and in which individual human beings become the central dynamic in planetary well-being.

Are we any nearer finding a route toward this goal?

Having accepted that not all innovative ideas come to fruition, and that much work is needed to transform an un-tried, innovative idea into workable reality, how much of this work has now been done? Does it still look like pie-in-the-sky, or is it beginning to get some flesh on its bones?

We've looked very inadequately at some trends in the current economic scene, drawing heavily on the work of experts in the field such as Peter Drucker, Robert Heller and Tom Peters, and come to the conclusion again that tinkering with an obsolete financial system will no longer do.

We've seen that capital flow has replaced market demand as the fly-wheel of today's economy.

We've looked at some of the games people play with money and at the wild fluctuations in exchange rates on the capital markets of the world.

We've seen how international corporations make a laughing stock of national governments in their attempts to regulate their internal economy.

We've seen, with Peters, that 'chaos is now the prevailing norm' in the business and financial world, and noted his observation that chaos can and must be used as the source of market advantage.

We've noted the similarity between this situation and that described by Ilya Prigogine with regard to the creative potential of chaos in living systems, taking heart from Prigogine's intimation that the emergence of a new structural and functional order actually seems to demand conditions of severe fluctuation and instability.

And we've suggested that rightly used, the current precarious situation of Planet Earth could be remedied by an equally flexible approach, using chaos not as a source of market advantage, but as a spring-board to wellbeing for mankind.

We remember Prigogine's observation that the more complex a system is, the more numerous are the types of fluctuations that threaten its stability, and his question as to how systems as complex as ecological or human organizations can exist. We remember especially his conclusion:

> where species and individuals interact in many different ways, diffusion and communication among various parts of the system are likely to be efficient.[1]

We've built our road to dynamic stability on this insight, concentrating first on diffusion through devolution: of political power to individuals and small local groups; of large-scale monoculture and speculative land ownership into smaller, individual, organic-type farms; of problems of industrial waste by reciprocal usage of resources; of sewage pollution through localized treatment and return to the land; and of fiscal resources through the distribution of public moneys to local bodies on a head-count basis.

We've insisted repeatedly that communication is the key to all these things, and that the technology already exists to integrate their random and multifaceted elements.

We've looked for the ways in which human beings most naturally interact and most easily communicate, noted that in nature infinite numbers of apparently random events arrange themselves in simple and beautiful overall patterns, noted the adverse effect of placing too much emphasis in a complex system on a single variable, and expressed the belief that the proper integration of apparently random human resources would create an equally satisfying overall order.

What remains to be done to give substance to my argument? A great deal, I'm afraid. Some of our most intractable problems haven't been tackled at all. The population explosion, for instance. And unemployment. The energy crisis. The vested interest of giant corporations. And perhaps most crucial of all: How, even in a world made ecologically aware, do you ensure that everyone votes on every question to the benefit of the whole organism? Finally comes the need to envisage more incisively the nature and structure of a global information network.

CHAPTER FORTY FIVE

9.7 The Population Explosion

Summary: In any living organism there is a genetically programmed number of cells beyond which the life-form ceases to grow.

With or without our conscious volition, one way or another our human numbers will be checked.

An exploration of ways and means.

The Population Explosion

To begin with, the daunting question of just too many people.

Valiant efforts have been made in a number of countries to control unmanageable numbers, and in some parts of the world, notably in Communist China, Draconian measures were tried for a while. Families were limited by decree to one child. The population growth-rate declined dramatically, but, as is now widely understood, such measures have unacceptable repercussions on the lives of individual human beings. Report has it that in China, while this legislation was in force, desperate parents delivered of a daughter while desiring a son, resorted to infanticide. I have no statistics on the matter, but clearly such measures are not the answer. Something has to be done, and quickly, lest we suffocate in the sheer weight of human numbers. Families are personal; the issues sensitive; interference of any kind understandably resented. So what do we do?

The current view, born out by falling birth rates in many Western countries, is that education, a higher standard of living, and reasonable financial security are the only humane and effective routes open to us. However, there is a serious ecological constraint to this argument, for even were it possible to give everyone immediate access to Western standards of living, pressures on the supply of fossil fuels and other resources could not sustain it; atmospheric and water pollution would rocket out of control; and the life-support systems of Earth would be destroyed.

Is there no way out? The one thing on which everyone seems to agree is that no amount of conservation of whatever kind can work while human numbers continue to soar at their present rate. It is a gloomy scene. And the choice is no longer entirely ours. Much destruction has already been done; much will boomerang, and many people will die. But at least we now know what the dangers are. We can go on from there.

In any living organism there is a genetically programmed number of cells beyond which the life-form ceases to grow. If this proves true for planet Earth, we must be very near to the critical point at which, by whatever mechanism, the apparently uncontrolled human birthrate will level out. We see this around us in all of nature. Where population outstrips resources, a falling birthrate seems to be achieved either by subtle physiological mechanisms, by the dramatic suicide of animals such as lemmings, by disease, or worst of all by famine. We humans have added to this rather frightening list pollution, global warming, destruction of the ozone layer and nuclear war. With or without our conscious volition, one way or another our numbers will be checked. Which way will it be?

It seems to me that perhaps some of the measures already discussed in this book, involving the diffusion of power, the more equitable division of land, the assumption of individual responsibility, the alleviation of poverty through local initiative, and the subsequent diversification of life-styles, may have beneficial effects here too. *It involves a man's perception of 'the good life'.* If everyone perceives the American-style, high-energy, pressurized, throw-away society as the peak of human attainment, we have problems. We've already observed, like others before us, that besides being ecologically disastrous, life at this scale and pace is by no means always happy. Drugs, suicides, ill-health and crime belie the vision of a contented community. We need to delve deeper than this. Perhaps the diffusion of congested populations into smaller diversified units, and the attainment of economic self-reliance discussed above might help, but it doesn't cover everything. If higher standards of living are the only acceptable route to a falling birthrate, what do we mean by 'high'? What rating, for instance, would we give to tribal peoples?

According to the human rights group, Survival International, there are some 200 million people worldwide who can be described as 'tribal'. Such people live in small cohesive communities which, given sufficient freedom of movement, are economically self-sufficient and are usually part of larger, culturally distinct groups, which may or may not be capable of acting as political units.

Tribal peoples, among whom are numbered hunters and gatherers, shifting cultivators, herders, subsistence farmers and fisher-folk, can

> also be defined by the absence of some of the characteristics common to state-based societies. They may not use money. They may have no systems of writing. Leadership patterns and decision-making processes are complex and power is diffuse. They may have deep-rooted traditions to ensure material equality, though not necessarily social equality, and they have strong ties to their traditional lands.[1]

It is common knowledge that these traditional lands, often sacred to the people who inhabit them, frequently lie in territory that can be profitably logged, mined, farmed, ranched, dammed or otherwise 'developed'. In consequence, tribal lands are increasingly being expropriated 'in the national interest' for development projects which, with hindsight, are often found to be to the detriment of the ecosystems involved. The tribal peoples themselves, their culture destroyed, having been forcibly dispossessed of their land and deprived of their traditional means of subsistence, usually end up poor, malnourished, disease-ridden and bedevilled by drugs - burdensome waifs on the edges of the very society which has created them. Constraints of a well established culture once removed, the birthrate can only soar unchecked. Could anything more graphically describe the dangers of trying to force traditionally diverse communities into a uniform state-controlled economy? As in nature, diversity of peoples is of the essence.

Over-population is calculated on the relationship between population numbers and the carrying capacity of the land, with methods of land usage very much part of the equation. Now tribal peoples are traditionally considered to be environmentally destructive, besides being centuries behind the more 'civilized'members of homo sapiens. Setting our sights on models for a good life, we have to ask ourselves if this is true? In Earth Report 2, Edward Goldsmith and Nicholas Hildyard find the reality very different indeed.

> Tribal peoples are poor only if poverty is identified with material deprivation. Certainly, many have few material goods, but then these are of little value to people who often lead a nomadic way of life. In fact, they do very well without them and there is no evidence that they feel in any way materialy deprived.
>
> On the other hand, their biological, social, spiritual and aesthetic needs ... are well satisfied. The diet of tribal peoples, consisting of a rich variety of fresh and uncontaminated foods, is generally excellent; their health status also seems very good; and their extended families and cohesive communities admirably satisfy their basic social needs.

As for their agricultural practices, time has proved these to be not only environmentally benign, but also sustainable.[2]

Not all of us would be happy with such a life, but it does give one food for thought. In contrast with our frenetic Western pursuit of material possessions, it has its merits. In many respects it is vastly superior to our own. One cannot avoid the conclusion that were it not for poverty, many would gladly return to traditional ways of life.

We have some valuable lessons to learn from tribal peoples. In the vexed context of population explosion, the key lies in their redemption from poverty. Tribal lifestyles married to a Western-type economy do not bring happiness. But we begin to discover that neither do ours. If family planning depends on the good life, as the experts proclaim, we have to move on to something new. And though for many help must come too late, a saner social and economic system might ensure that nature's more drastic measures be partly forestalled. The new measures would certainly help to regenerate diversity, giving rise in the process to a more contented community. A new orientation of family life might be expected to follow.

CHAPTER FORTY SIX

9.8 Unemployment and the Work Ethic

Summary: The deprivation experienced by people with no work, no motivation, and no opportunity to use or extend themselves, can hardly be imagined by those whose work fills their lives.

Over-emphasis on the 'Symbol Economy' of capital flow and international monetary fluctuation is wrongly being allowed to dictate employment.

Unemployment And The Work Ethic

The next most intractable problem facing the capitalist world would seem to be unemployment. I confess to impatience, if not to fury at this one. So much needs to be done, yet because we lack the means to pay for their labour, despairing people sit idle.

Statistics vary from nation to nation, but the millions unwillingly out of work grow daily. And people starve. It is a ferocious insanity.

How have we gone so terribly, tragically wrong?

The basic question in the field of manpower is not: How can full or maximum employment be achieved and maintained without prejudice to profits?

It is: How can the needs of planet and people be met in such a way as to ensure health and productivity for the planet, and realization of the full, happy potential of its people?

Happiness is mankind's natural state. In our pursuit of money we have forgotten this. We must get back to it. If Gaia is lucky enough to reach maturity, this is how it will be.

Speaking in a very different context, Drucker observes that organized abandonment of the obsolete is the one sure way for society to focus the vision and energies of its people on innovation.[1] We know that the full usage of human potential in self-chosen paths of productive activity automatically ensures employment. So why let money get in the way? It is the means to an end, not an end in itself. If it no longer does its job for the majority of the population, get rid of it. We should act on this.

Once again what is needed is widely understood priorities concerning healthy environmental norms and their relation to human wellbeing, so that people, properly informed, will choose those areas of activity which are seen to be in the longterm interest of themselves and their families. For the first time we have the capacity to keep ourselves informed. The choice of activity can now be up to us.

This means that when the drive for profit and the struggle for a living wage no longer apply, nothing will stand in the way of useful activity for everyone, save long submission to outdated norms. In economics as well as in ethics, 'God fulfils himself in many ways, lest one good custom should corrupt the world.'[2]

The extent to which world finance, outside their own control, shatters the lives of humble people, is well illustrated by the plight of the people of Ngwavuma, Natal/Kwazulu.

'News', official mouthpiece of South Africa's 'Operation Hunger', carries the story in its April 1991 edition.

> Urgent intervention to 6 000 children is necessary....
>
> Deputy Director Mpho Mashinini reported that when the Executive Director and he had visited the area a few weeks before, their reaction was one of total disbelief, since they had not seen such visible child malnutrition since 1986.[3]

Children had become the tragic victims of massive retrenchments in the mining industry. Why the retrenchments? Because the price of gold on international markets had dropped to a level where many Transvaal mines were no longer viable. This drop, combined with rising costs, meant imminent closure of a number of mines, and retrenchment in migrant labour. Why had the price of gold dropped? Reduced to its most simplistic terms: Because an end to the Gulf War restored investor faith in the American dollar. Because panic in the business world sends the gold price soaring, while confidence makes it drop. Dollar up, Gold down.

But why should international market fluctuations and the price of gold affect a poverty-stricken black family in rural Natal/Kwazulu? People buy gold as a hedge against uncertainty. When things improve they buy less, and the price comes down. It all sounds very reasonable. But many thousands of miles away, as a result of events beyond their control, outside their experience, remote from their lives, and beyond their understanding, children die. Must even one child die for reasons like these?

When thousands of men streamed home to Ngwavuma from the Transvaal gold mines during the period 1989/90, - after a loss of 80, 000 jobs in the industry in 18 months - young children in this remote corner of rural Zululand suffered from malnutrition as never before. Fathers could no longer support their families. Few would find work elsewhere. But given a change in the way we think, need this mean starvation? I really think not. We have to break the link between wages and work. And the key is a local one. It lies in local development and in self-help. It lies in communication with the outside world. It lies in people's willingness to help each other.

There is much that needs doing in Kwazulu's beautiful hills. The country has had its share of drought and flood. Large areas are badly eroded. Others stripped of trees. Where once the verdant earth was watered by springs and streams, much is now bare. Water now has to be carried long distances. People need homes and children need schools. But the people are 'unemployed', and children die. And this in spite of thousands of idle hands.

Unemployment in this area, because of its heavy reliance on mine salaries, is estimated to have risen from 65 percent in June 1980 to over 90 per cent early in 1991.[3] On a recent visit to the area, Operation Hunger's executive director Ina Perlman described conditions as the worst in the country. Drought and an influx of Mozambican refugees worsened the situation. Ingwavuma-based doctors said that malnutrition had become the the worst killer of all. The area reported a 400% increase in malnutrition admissions to hospital between January and December 1990. Famished children were begging at the roadsides, stealing from school stores, and eating leaves. Entire families were scavanging and surviving off wild plants.

In situations like this public concern and donation helps. So does the channelling of Government funds. So do grants in aid from abroad. So, most of all, does the work of bodies such as Operation Hunger, which not only try to feed growing numbers of desperate people, but try to assist them in learning to help themselves.

Crucial to success is getting to know the local people . Moving among them, listening, teaching new skills, finding markets, boring for water, or starting up vegetable plots, Operation Hunger's indomitable director leads and inspires a growing team of helpers from the communities themselves, finding out first what they would like to do. And this insistence that wealth lies in the value of each human being's own initiative, is really the crux of the matter. We must find ways to help people help themselves, which, as things stand at present, requires the injection of massive amounts of money outside the community itself. It is a despairing foolishness, but one we can change. With self-improving work crying out to be done, and people ready and eager to do it if given a little outside help, no-one should have to be hungry or deprived.

The deprivation experienced by people with no work, no motivation, and no opportunity to use or extend themselves, can hardly be imagined by those whose work fills their lives. Without motivation and hope the spirit dies. It would even be true to say that effective work is motivation in action. At its most useful it engages a man's full capacity, extends his potential, and serves, directly or indirectly, those he most loves. Absorbed and intense in his work, he is at his happiest. And because he lives always at the very edge of his capacity, creative forces come into play. His growth is enhanced and his spirit enriched by labour.

We have insisted throughout this book that the human race is entering upon a new phase of evolutionary consciousness in which, by converging on itself through communication and spiritual communion, it is destined to make a qualitative leap into a wider form of transcendental consciousness. The enrichment by full participation of every man, woman and child on the planet is therefore paramount. Situations like Ngwavuma are endemic to Africa, but lately spring up like mushrooms all over the world. They are a symptom of much deeper ills. We dare not, and need not, let them go on.

We used the price of gold to illustrate how far our economics have strayed from their legitimate job, how totally divorced from the needs of people, how poor at circulating so-called wealth. Above all to illustrate the total inability of a 'symbol economy' to handle the conditions of unemployment that spell starvation to so many of the world's poor. It simply will not do.

Perhaps it is necessary to stress again at this point that this is not an attack on the capitalist system per se. Even less is it a plea for communism, socialism, or any form of state control. State control of the economy is the last thing we want to see. It is even less efficient than our own. What I believe we need, is that the notion of market demand be carried to its logical conclusion of producing, growing and interchanging our legitimate requirements and desires without the distortion caused by the monetary system. I believe it is time for the notion of 'work' to be divorced totally from the notion of 'wages'; for all beneficial activity, including study, home-making, the rearing of children, and the host of other necessary activities presently outside the monetary economy, to be embraced in the concept of useful work; and for any form of useful work to entitle men and women to a reasonable share in the resources of the community. For the very first time in the last three hundred years, with electronic networks moving into place, we are in a position to move towards this goal. It all adds up quite simply to meaningful participation in the community.

I am sometimes asked why, if food is flowing in anyway, a man should bother to work. 'Won't the willing horse be supporting the sluggard?' We're looking here at the question of motivation. Progressive managers solve this problem by rewarding increased production. Reward is given to the appropriate team. Knowing the power of peer approval, the team is then left to divide it among themselves. They know who the workers and the loafers are.

In industry the reward is often a share of profits. But as we move to a moneyless economy, what about a share of *benefits*, awarded to the team as before, and shared out by them, but consisting of things like occupation of the first house a group builds; or ownership of the first available car, or a site on the favourite bit of river; or the first available trip overseas? Why only money?

And what about leisure? Do our young people really *want* to work the way the modern young executive works, with no time left over for living, and a heart attack in the pipeline a few years on? Supposing increased productivity were rewarded with shorter hours, or a holiday? - for the group if they'd made enough of what was needed (we'd know under the new system how much this was) - or for the individual in recognition of outstanding contribution. Individuals or groups could disperse to do the things they best liked doing, individually or together, from pottering about with junior to designing a space ship - because they had earned that most precious commodity *time*. Eventually, as the system really moved into gear, most people would have migrated to their favourite kind of work anyway, and would be working for the sheer love of it.

Motivation In The Work Place

This brings us to the question of motivation in a more general sense. In a moneyless economy, would there be adequate motivation to keep society operable?

For most people just working for the benefit of the family they love is motive enough. But there are a number of others. Granted that the Mother Theresas of this world are few and far between, there are many ordinary dedicated and public-spirited people already engaged in helpful activity, both paid and unpaid. And there are those who are perhaps paid, but who really love what they are doing and wouldn't be parted from it.

- There are the scientists, filled with insatiable need to know.

- There are the sportsmen, searching for excellence, and, for some, the added spice of danger.

- There is the passion for music or art or making things with one's hands.

- There are engineers and architects who invent and design and create, whose creative faculty is life to them.

- There's the teacher's interest in seeing a child's mind grow.

- The farmer's love of the land and the beasts of the field.

- The true scholar's passion for learning.

- The gardener's care for the smallest plant he tends.

- And there are those who love the wilderness.

All these are motivation enough, and the list is endless. But what about those who haven't any such love? And what of the really unpleasant jobs, that nobody cares to do - jobs that only get done to earn a wage?

Many of these, I think, will already have fallen away. Waste materials will be re-used; sewage will be more easily processed. Dangerous and distasteful work will largely be done by machine. So the problem is motivation for the un-inspired.

So far as the labour force is concerned, industry has led the way in this. In spite of continuance of the concept of 'money'in return for labour, and in spite of the paralising strikes for more of it, it is now well understood that worker participation means a great deal more than wages alone can supply. We've been rather slow in coming to this, but in the last thirty years, meaningful participation of workers in industry has at last been widely explored.

Some of the early experimental work was done in the 1960s by a Norwegian psychologist named Einar Thorsrud, fortunate enough to obtain the co-operation of members of the like-minded Tavistock Group from London. Reporting on the first phase of their investigation, Thorsrud concluded that it was necessary to begin on a level where the majority of the workers had both the ability and the willingness to participate. 'In our opinion, that means that we must begin at the individual's work situation and improve conditions for personal participation there'. [4] He postulated six basic needs that an individual seeks, to satisfy him in his work: 'a real content, including possibilities for variation; continuous learning opportunities; the chance to make decisions; recognition and respect; an understanding of the connection between the job and the rest of the world; and a view of the job as a step toward a desirable future.'[5]

Thorsrud explored the possibility of achieving these conditions by obtaining the co-operation of a number of enterprising firms prepared to try out what were then quite startlingly new ideas. Chief among the principles applied by the Thorsrud group was the notion of 'autonomous groups', which, when properly established in manufacturing plants, 'develop a powerful dynamism and momentum of their own in harnessing the latent abilities and ideas of workers.'[6] In spite of some impressive results, not all the Norwegian experiments were successful, but the idea of individual participation in decision-making at all levels had come to stay. Subsequent work in Yugoslavia, West Germany, Britain, the United States, Japan and others all show variations of the same central theme: freedom to develop one's best potential breeds responsibility; combined with group identity in a team, individual involvement does result in impressive quality control and higher productivity.

Later decades added the concept that workers might own shares in the company, so that a man was to some extent working for himself.

These progressive ideas developed in an era of economic prosperity in which growth seemed the only legitimate aim, when industry seemed able to absorb almost unlimited labour, when education had resulted in high expectations, and when people were no longer so poor that they felt compelled to put up with any structure employment imposed.

The difference in today's situation needs no comment. But while the basic principles of individual involvement remain, and industries vie with each-other in progressive management practices, a man's initiative is used - as it always has been - essentially to promote market advantage. Industry can neither absorb the millions of displaced people streaming in to the towns, nor guarantee employment to those unfortunates made redundant by economic recession and political unrest.

At its best, the Western system was evolved to provide wealth for some, and a 'trickle-down' prosperity to all levels of society. But the system has become distorted in a number of ways, not least by the frightening number of people it can no longer support. It simply does not work any more. Nor, I believe, can any other system imposed from above. Hierarchical structures are as hopelessly outdated in the business world as they are in politics. A typical scene today, in Europe, in Africa or indeed almost anywhere else, is of thousands of jobless people marching through the cities of the world in protest, demanding of their governments what neither government nor society can any longer provide - the security of a paid job. But there have been pointers to a better way, notably in the work of the late E.F. Schumacher, himself an economist, whose work was discussed at some length earlier in this book (Chapter 23).

With the possible exception of Ghandi, no-one knew better than Schumacher the therapeutic value of work, nor felt more deeply for the poor than he. Writing nearly twenty years ago with a courage and vision way beyond his time, he penned the now famous words: 'If [economics] cannot get beyond its vast abstractions ... and make contact with the human realities of poverty, frustration, alienation, despair, breakdown, crime, escapism, stress, congestion, ugliness and spiritual death, then let us scrap economics and start afresh.'[7]

Schumacher's work justly received the fame it deserved, but the means to 'scrapping economics'were not yet fully at hand. They are so today. The road to satisfying human labour lies in the very processes of diffusion and communication we've seen to be central to so many of man's activities - in industry, in agriculture, and in politics. It is now well understood that 'people can be themselves only in small comprehensible groups', so we have to de-centralize and diversify. Both in business and in agriculture we have to develop articulated systems composed of millions of individual people in a multiplicity of localized small-scale units, in which geographical and local cultures ensure the colourful variety of occupationand interest that nature so abundantly displays. And the means to integrate all these informal, individualized activities lies in the technical capabilities of Computers and Communication - a pseudo, fine-grained, articulated nervous system encircling the globe. We'll come back to this. Given wider access to land, freedom of initiative, communication, and the availability of appropriate knowledge, all that remains between a man and enriching work lies in an outworn system which we are now in a position to replace.

CHAPTER FORTY SEVEN

9.9 Groups

Summary: Computers make possible the integration of human beings in committed, flexible, multi-skilled, self-managing teams that reflect in society the congenial relationships of living organisms.

Groups

Much emphasis has been placed in this discussion on the individual and his need for *Lebensraum*. But individual needs are of course only one side of the coin of social well-being. No man operates in isolation, no matter how exclusive the niche he may have created for himself. The whole history of cultures has evolved round man's penchant for interactive relationships. Let you spread mankind as evenly across the planet as Arctic snow, and in no time at all he will be found in close, congenial little groups, building all manner of relationships, competing, interacting, communicating. Moreover some of his more ambitious projects can only be done by large numbers of people working in concert. So the question arises: Starting from where we are now, what form is this multiplicity of small-scale units going to take, and how can they best evolve from existing economic structures?

Peters speaks of the 'Winning Look' as comprising 'customer-oriented specialization and more moderately sized businesses, oriented toward differentiation, niche creation, quality-consciousness, service-consciousness, flexibility, speed of innovation, and the use of highly trained, flexible people as the principal means of adding value.'[1] All this is perfect for the new, information-based organic order, and when he adds that 'All this adds up to a requirement, not for structural or procedural tinkering, but for a revolution in organization: more autonomous units, guided by a *coherent vision* rather than by memorandums and managers-as-cops, and manned by involved workers with a big stake in the action and hell-bent upon constant improvement[2] it could have been tailor-made.

Of particular significance in the organic context is that the trends demanding priority in today's business world prove to be the very same as those required for people and the planet as a whole. Peters insists that new attitudes are also required - 'especially commitment on the part of managers to the idea that suppliers, workers, unions, distributors, and customers are all partners in the common endeavour'.[3] Exactly so. But the crux of the matter lies in the nature of our vision and the goal of the 'common endeavour.'

Quite the most dominant group structure of present-day economics is the corporation, collectively 'owned'by shareholders, and theoretically open to investment by anyone. At first sight this would seem a highly desirable development, but the truth is very different. Since shares are only within the reach of people with capital to invest, there is no direct way in which the poor can benefit from corporate ownership. Moreover capital is usually provided by very large numbers of people who have nothing to do with the actual running of the corporation, nor with decision-making within it. These functions are left exclusively to management. In consequence shareholders have until recently exerted little moral restraint on the activities of the corporation, and still have no liability for the conduct of its business. Their single direct interest has been that the corporation should show consistent and increasing profits, a situation often in conflict with the interests of employees, and with ecological concerns in particular. The larger corporations thus constitute awesome power-centres. In practice they have become autonomous units which largely dictate government policy through manipulation of the national economy, while cross-border activities set them beyond the control of any government.

But is big business the only social institution to wield such inordinate power? Drucker notes that we have become a society of organizations. In the text-book doctrine of democracy still taught in our universities, the system is so organized that power is finally vested in one central government. 'But the society of organizations is a pluralist society. In open defiance of the prevailing dogma, it contains a diversity of organizations and power centers.[4] The business enterprise is only one. Drucker mentions labour unions, farm organizations, health-care institutions, schools and universities and the media, and observes that 'even government is increasingly becoming a pluralist congeries of near-autonomous power centers, very different indeed from the branches of government of the American Constitution.[5]

While all these groupings have an explicit or implied social intent, all, with the possible exception of State schools and the universities, embody profit-making as their prime motivation. The privatization in many countries of health services has ensured their inclusion in this ever-widening and rather frightening generalization. But equally noteworthy is the proliferation of enormous numbers of highly diverse single-purpose institutions whose motivation has nothing to do with economic priorities. They range from the protection of minority rights (including those of children, animals, the physically disabled or the homosexual) to religious groupings, preservation of the environment, the care of refugees, study of cultures, protection of national or historical monuments, care of the terminally ill, sea or mountain rescue, and a host of other dedicated enthusiasms that grow daily more numerous as like-minded people gather in small identifiable and interacting groups in the interests of society as a whole.

As with business concerns, the ubiquity of the new electronic media ensures that many of these groupings begin to have cross-border affiliations, and to link together individual people of many and various cultures and political persuasions. It is not an exaggeration to say that this building of personal relationships, both in business and in voluntary group activity, is the most significant offshoot of the new technologies. Linking people together is *the* function of computer power.

It also brings together institutions of widely differing character and aims, enabling people with an interest in both organizations to work together to the benefit of each. Such a partnership in South Africa is the recently established 'Green Trust', in which Nedbank and the Capetown based Southern African Nature Foundation have teamed up, to further the overlapping aims of both.

This partnership has embarked on a far-seeing new venture embodied in a 'Green Trust Credit Card', made highly visible by the internationally known WWF (SANF) logo and an attractive wildlife painting in colour on the face of the card.

Nedbank has started the project going with a R5 million donation over five years. Thereafter 1% of every purchase made on the Green Card goes to the Green Trust as a donation from the bank, to be channelled into a variety of conservation projects. These include integrating the needs of the local people with the protection of landforms and ecosystems, provision of information on such vital issues as the management of water resources, maintenance of genetic, species and habitat diversity in flora and fauna, care of marine and coastal resources and wetlands, environmental education, and research.

This is a splendid example of the power of big business to stimulate interest in ecological concerns, as also in the new trend of major corporations to take a lead in promoting them. And of course computer power is the key factor in the enterprise. Simply by using the computer-controlled 'Green Trust Credit Card', individual people with like vision but limited means are enabled to cooperate in large-scale undertakings, and the well-being of the whole area is enhanced.

The integration of individual people and groups through electronic media takes many and diverse forms. As new technologies emerge from the laboratories and become available to the public at large, people are increasingly enabled to enmesh their thinking, their interests, their hopes and their activities. And the new machines play a remarkably humanizing role.

Quoting author Howard Rheingold in *The Media Lab*, Stewart Brand notes, for instance, that the most surprising and consistent quality in 'e-mail' communities is the human warmth they develop. 'They are a form of conversation. 'There's the permanent temptation to log in: There's always another mind there ... It's like having the corner bar, complete with old buddies and delightful newcomers and new tools waiting to take home. ... It's a place.'[6]

The key, then for effective work in any useful sphere is that people with a common interest get together in small, self-managing teams to attain a specific goal. Their self-sufficiency is made possible by the decentralized availability of information, and by local concentration of the physical assets required. Why do they work so well? Peters puts it this way: 'Quite simply, people or groups of ten to thirty can get to know one another well, can learn virtually every one else's tasks, can be gotten together with little fuss, and under enlightened leadership can readily achieve unit cohesion and esprit.'[7]

So the goal is either small, self-sufficient units, or, within the established giants of big business, the 'small-within-big' principle, engaging maximum participation by each and every individual in a coherent team. Peters notes that committed, flexible, multi-skilled, constantly retrained people, joined together in self-managing teams, are the only possible implementers of successful business strategy.[8] As a recipe for niche creation in the context of a living world, it could hardly be bettered.

CHAPTER FORTY EIGHT

9.10 Energy

Summary: Three main aspects of energy use have to be addressed before we can apply living principles in this field. These are:

- The development of diversity

- The saving of energy

- The up-grading of technical efficiency

Thereafter energy use can begin to operate on an intricate and delicately balanced system of energy conversion based on information.

Energy

a.) Principles

Have I anything new to add to the vast specialist literature available on energy? Do I even dare to try?

Only, I think, to suggest that the same principles apply in the field of energy as in everything else we've discussed so far. Diffusion in the use of resources, local diversity, individual initiative, diversity of life styles, niche creation, local self sufficiency, reciprocal energy use, and abandonment of the idea that a Western-style high-energy, high-consumption, throwaway society is the peak of human attainment. We've already noted that living systems do not operate on energy in the physicist's meaning of the term, but on an intricate and delicately balanced system of energy conversion based on information. We have, I think, to learn to do the same. But before we can make such a switch, there are three main aspects of energy use that we have to address. The first is *development of diversity*, to include a multiplicity of renewable-energy sources; the second lies in the actual *saving of available energy*; the third is to upgrade *technical efficiency* in energy use. All are closely interlinked, and all are essential to global integration.

It all sounds pretty straightforward, yet there are serious economic and political objections to this kind of thinking. Elkington, writing in 1984, points out that there are many who see renewable-energy technologies as intrinsically revolutionary, 'facilitating the break-up of highly centralized societies, bound together with high-voltage energy grids and supply lines tapping into the coal, uranium and oil-fields of the Middle East and elsewhere'.[1] The arguments for survival put forward by such research groups as the New Alchemists, on the other hand, envisage an increasingly decentralized settlement pattern, 'tapping into renewable-energy flows and, in the process, dispensing with the services of the conventional electricity supply industry.'[2] It is one of the strengths of solar and related soft-energy paths that they potentially give a high degree of self-determination to local communities and different regions of the country, and power blocs, whether economic or governmental, are understandably anxious.

They fight a losing battle. In the seven years since Elkington's book was published, it has become abundantly clear in many parts of the world, that highly centralized societies are on the way out. The Baltic States demanded cessation from the Soviet Union, the Tamils want independence from Sri Lanka, Sikh separatists from India, Quebec from Canada, Basque separatists from Spain, Croatians from Serbs, and so on. It is an interesting paradox that the closer we draw to a technically feasible integration into one world, the more individualistic - and the more vocal - do splinter groups, ethnic or religious groupings, and small members of conglomerate nations become in their demand for autonomy. Even in small countries, local communities tend more and more to demand control of their own affairs. Arguments for centralized control bolstered by high-energy grids and massive corporate economic institutions are thus out of line with the political and economic realities of the 1990s.

Were central governments and corporate power still able to 'deliver the goods', and to do so for the mushrooming millions of exploding populations at a price they could afford, things might be different. They are clearly not able to do this. Even the United States of America, the richest and freest nation the world has ever known, is unable to offer all its citizens a life-style comparable to that of the privileged few, and this in spite of the fact that America consumes an unacceptable percentage of the world's available energy.

Add to these trends the requirements for ecological survival of the planet, and it becomes very clear that in energy patterns as well as in economics, the future has caught up with us. Present patterns simply cannot go on. Like it or not, we have to develop diversity.

But can governments and big business not keep control of alternative energy sources as they have with oil and coal? The prospect of local communities declaring independence from utilities and energy corporations once they have developed their own renewable resources, is unlikely to prove popular with the corporations involved. And there does seem to have been evidence, particularly in the US, of collaboration between governments and corporate power to maintain control.[3] According to Ray Reece, the technique envisaged was 'to emphasize those applications of solar energy that "are most compatible with the present system of capital-intensive, centralized power facilities".[4]

One must remember that renewable energy sources cover a very wide range of options, from simple small-scale devices to very high-tech capital-intensive installations. One thinks most naturally of direct solar power, but wind, wave, and ocean thermal layers are all products of solar energy, as are plants, algae, and the bacteria responsible for fermentation. Moreover 'hard' and 'soft' energy paths are not mutually exclusive. High-tech installations such as those of 'power tower'photovoltaic cells, ocean thermal generating plants and giant, expensive wind turbines are ideal tools for meshing the burgeoning solar market into the larger corporate power structure.

Reece believes that 'the rapid deployment of renewable energy technologies - controlled by people and institutions at the local level - could trigger a peaceful revolution.[5] In this view 'the 80-year trend towards concentration of wealth and power in the hands of the corporations could begin to reverse. The drift toward coal and nuclear power as replacements for oil and natural gas could be halted.'

> A revolution emphazing local self-reliance and social equity instead of servitude, waste and mega-profits would, of course, undermine the corporations and financial institutions that presently dominate American life. [6]

The corporations, Reece suggested, do not intend to let this happen.

We've already noted the growing inability of the corporate state to 'deliver the goods'to broad masses of its people, or to provide them with steady employment. We've noted the determination of increasing numbers of angry people to put an end to these imbalances (even to the extent of plunging entire nations into chaos). And we've emphasized continually how very-large-scale single-purposed human activities tend to brutalize the environment. All these factors point to the development of smaller-scale, independent and local solutions in energy as in so many other endeavours. But there is another, purely economic reason why small energy plants are likely to flourish - the lively international competition for ever shrinking markets, as more and more previously underdeveloped countries begin to invade the export field. Countries like South Korea or Taiwan are unlikely to ignore so fertile a field as small-scale solar systems for their burgeoning export industries. While big oil companies dominate the high-tech, high-energy photovoltaics industry at the present time, companies like Japan's Sharp and Matsushita, already busy in the solar field, are unlikely to undervalue the enormous potential market for small-scale plants. Elkington notes that US oil giants are likely to be outflanked by the Japanese in this field, just as they are being outmanoeuvred in the microchip market.

By and large then, whatever the understandable reluctance of big business to change the status quo, and whatever their success in channeling solar power into high-energy, high-capital, high-tech enterprises, the fact remains that the heat of the sun is naturally decentralized. In the remoter regions small-scale solar systems are already a viable option for radio telephones, navigational aids and the like, while lights, pumps, televisions and fridges are increasingly run on a variety of small-scale solar installations where mains power is not available. Both 'hard' and 'soft' paths will undoubtedly be followed in the use of solar energy for as long as the present economic system can be maintained. Moreover the highly evolved management skills of big business will surely have a significant part to play in the global integration of the future. But usage friendly to the environment and the overwhelming mass of its people will finally come about by the internal dynamics of business itself.

b.) The Development Of Diversity

Oil

If anything had been needed to underline the urgency of the energy crisis, it could hardly have been more graphically demonstrated than by the Gulf War - a war basically concerned with control of the world's oil supplies, and perhaps the most destructive and ecologically disastrous conflict in the history of man. For good to emerge, then, from anything so horrific, it must surely entail grave attention to the question of energy sources alternative to oil, in despite of the formidable vested interests involved. And following the Chernobyl disaster, the same must be said of nuclear power.

No one needs to be told that had a fraction of the money spent on the Gulf War been spent on the development of alternative energy sources, there would have been no need for a war. The vast international concerns who supply military hardware to both sides would have shown less profit. Many people involved in armament production might have lost their jobs. Perhaps whole economies would have staggered and cracked. But when the final sums were done and the costs assessed in human suffering, would humanity as a whole have been the poorer?

And what of research? A staggering percentage of high-tech research stems from the needs of war. We still insist on calling this 'defence'. We've been conditioned to go on thinking this way, long after it ceased to be useful. But the time has come to use our powers more creatively, and we can - perhaps nowhere more usefully than in the energy field. So what do we do?

It can be argued that we've only just scratched the surface of global oil reserves, rendering invalid any talk of limited resources. But it isn't only quantity involved in this. It is convenience, accessibility, cost, and danger. It is the ever-present threat to marine life of large-scale oil spills; the enormous consumption of energy involved in shipping oil around the world; and the pollution caused by large-scale burning of oil in power-generation and industry; all this even without a war. Most wasteful of all, perhaps, is that oil represents the stored solar energy of life-forms that existed millions of years ago, that could be more suitably used. Brian Ford, in his book *Microbe Power*, has an interesting comment on this:

> [Each] tonne of oil burned is irredeemable loss. I have long believed that in principle it is no better than burning banknotes for the sake of some temporary warmth, for oil contains so many valuable chemical raw materials that *burning seems the last thing one should do with it*. Further, plants alive today can produce captive solar energy for us.

If we accept that oil - the life-blood of heavy industry - is now dangerously polluting, expensive, and inadequately used, what are our other options?

Nuclear Power

The dream of safe, clean, economically viable and inexaustible supplies of energy from nuclear fission has lately been debunked, and in the developed countries new installations have largely been stopped. In the US, no new plants have been ordered since 1978, while 108 have been cancelled since 1974.[7] But we have yet to dispose, without polluting the earth or sea, of vast quantities of radioactive waste from those plants already in use. We haven't the slightest idea how to do this. Nor are we yet able to assess how much damage has already been done. We do know that a year after the Chernobyl disaster, in the UK alone (2,000 miles from the accident,) 465 farms and 250,000 sheep in Cumbria and North Wales remained affected.[8] Whether in deaths from cancer or in damage to the environment, the long-term effects of radioactivity remain incalculable. So, for the moment at least, nuclear fission must be considered too dangerous to pursue.

The whole issue is immeasurably complicated by the vast amounts of capital involved, in the oil industry, in coal-fired power stations, and in nuclear generation of electricity. Big business can be expected to maximize profits on the remaining oil, while consolidating its already considerable investment in electrical power grids based on coal and nuclear energy. In present economic terms, this makes excellent sense. As a global issue it spells suicide.

Is nuclear fusion a viable option? Perhaps. If so, it lies in a future to which, so grave are our present problems, we may not have access until it is too late.

Coal

We have, for the moment, abundant supplies of coal. Useful indeed, but highly polluting, dangerous to human health, a major contributor to the Greenhouse Effect, and the prime cause of acid rain.

Until we start applying strict anti-pollution measures to the burning of coal, especially for the generation of power for countrywide electrical grids, it remains an unacceptable hazard.

It is a sad reflection on our misplaced priorities that the technology already exists to deal with emissions of sulphur dioxide and oxides of nitrogen from coal-fired power stations. The stumbling block is, as always, cost. But while sound economics demands a profit to stimulate growth, the cost to human life has become too high. We need to look with new eyes at our outworn memes, and search for new ways. Coal must be safely burned, or it must not be burned at all. And now that we know of alternatives, they should be more fully and earnestly explored.

Alternatives to oil and coal

There are, in our present state of knowledge, a number of useful alternatives, none of them adequate by themselves, but more than enough together for sensible use. Sun, wind and water all offer attractive options, ranging from small-scale local generation and use, to much wider, capital-intensive, high-tech development. Some of these will be discussed in due course in their relation to systemic principles. They entail the integration of a very wide diversity of energy resources and of the geographical distribution of natural energy flows, a match between the scale of power-generation and the type of end-use that would be involved, and sensitivity to the traditional or chosen life-style of the various local communities.

c.) The Saving Of Energy

Next to the idiotic squandering of precious resources in the manufacture of armaments, in war itself, and in its aftermath, energy wastage would seem to be most directly linked to the widespread practice of growing and transporting tons of oil, fertilizers, heavy machinery, pesticides, cash-crops and toxic wastes around the globe, and in using a 'hard-energy' path where little energy is in fact required. Maximum development of local energy resources, the local growing of foodstuffs for local consumption, a return to millions of small-scale organic-type farms and small-holdings are obvious means of counteracting some of these ills. Widespread organic farming practices would lead to a reduction in the use of artificial fertilizers, pesticides and heavy machinery, with a corresponding reduction in energy consumption. Maximum local production would reduce the transportation of foodstuffs, cutting energy consumption yet further. In addition the use of waste heat from industrial and power plants, recycling of 'throw-away' goods and a reduction in packaging would all mount up to a formidable reduction in the consumption of energy.

A corollary of these measures is to match energy resources and generation to the requirements of the job in hand. Amory Lovins argues[9] that a hard energy path is wasteful because the laws of physics require, broadly speaking, that a power station change *three units of fuel into two units of almost useless waste heat* plus *one unit* of electricity. These staggering figures mean that though electricity can do more difficult work than can the original fuel, unless it is employed in the performance of such work, the costly process of upgrading fuel is extremely wasteful. Industrial nations, accustomed to having electricity available for the most mundane tasks at the touch of a button, are using inordinate quantities of premium fuel 'for which their high-energy qualities are superfluous'.

> Instead, it was argued, where we want to create temperature differences of only tens of degrees, we should meet the need with energy sources whose potential is tens of degrees, not with a flame temperature of thousands of degrees or a nuclear reaction whose temperature equivalent is almost off the scale.[10]

This, Lovins concluded in a memorable phrase, is rather like 'cutting butter with a chainsaw'. One example of inappropriate application is the sole use of electricity to heat homes and bath water, when appropriate design and suitable building materials can reduce energy needs by the use of solar power.

Sunshine arrives on Earth at a relatively low temperature, making it highly suitable for tasks requiring low-quality energy.

Increasing The Efficiency Ratio Of Energy Usage

This aspect of the energy problem becomes startlingly visible when one considers that about 60% of the energy available in the coal, gas, oil and uranium we use to generate electricity is wasted before it is even put to useful work. 'According to the International Energy Agency,' say Goldsmith and Hildyard,[11] 'energy use in the industrialized countries could be cut by 30% by the year 2000 were the conservation measures that are (even now) economically viable to be implemented'. In the USA alone, according to Amory Lovins of the Rocky Mountain Institute, the potential for saving electricity through greater energy efficiency is 'equivalent to 50 Chernobyl-sized reactors'.[12]

In Britain, it has been suggested that the demand for electricity could be reduced by 70% if the most energy efficient lights, motors and appliances were to be used.

> Despite the cost-effectiveness of energy efficiency, however, there has been little enthusiasm for energy conservation programmes by governments internationally. Between 1972 and 1980, more than 90 per cent of the aid supplies to the Third World for energy programmes went to large scale electricity generation projects. Less than 1% was devoted to efficiency.[13]

Here once again we are up against political attitudes such as those evident in the Third World with respect to the 'alternative technology' advocated by the late E.F. Schumacher. It is a great sadness, but understandable, for historically the West has helped to develop the Third World with one hand, while milking it dry with the other. Perhaps the fast-changing attitudes of the industrialized nations themselves will spill over into the Third World as it becomes increasingly obvious that we are all in this together. The expertise that we need is already there.

A second aspect of efficiency in energy use lies in the changing nature of industry. Industrial production is switching dramatically from heavily material-intensive to far less material-intensive products and processes, due in large part to the emergence of the new high-tech industries.

> The raw materials in a semiconductor microchip account for 1 to 3 percent; in an automobile their share is 40 percent; and in pots and pans, 60 percent. But the same scaling down of raw-material needs goes on in old industries, and with respect to old products as well as new ones. *Fifty to one hundred pounds of fiberglass cable transmits as many telephone messages as does one ton of copper wire, if not more.*[14]

Drucker in *The Frontiers of Management* says that this steady drop in the raw-material intensity of manufacturing processes and manufacturing products extends to energy as well, and especially to petroleum.

> To produce one hundred pounds of fiberglass cable requires no more than *one-twentieth of the energy* needed to mine and smelt enough copper ore to produce one ton of copper and then draw it out into copper wire. Similarly plastics, which are increasingly replacing steel in automobile bodies, represent a raw-materials cost, including energy, of less than half that of steel.

This capacity to do more with less is characteristic of the technological age in which we live, and though many thinkers paint distressing pictures of a future world dominated by the machine, the extent to which increasing efficiency can assist in reducing ecological pressures is not always appreciated.

e.) Systemic Principles In Energy Use

One square inch of green leaf absorbs one calorie of energy from sunlight per minute. ' This may seem to be only a tiny bit of energy,'says Rutherford Platt, 'but when you multiply minutes into daylight hours and realize that a single maple tree may spread half an acre of leaf surface with several hundred thousand leaves, the input of energy into the leaf factories of a forest is incalculable.'[15]

Now the most interesting thing about about this kind of energy collection is the economy with which green leaves operate to perform photosynthesis,and the multi-purpose character of the end product. It takes four electron volts - or four photons - to tear apart the hydrogen and oxygen in one molecule of water. 'This precisely metered, this finely guaged act, is the fulcrum of life.'[16]

Drexler asserts that existing solar cells are already more efficient than plants,[17] a statement I find difficult to understand, for photovoltaic cells would seem to me to be doing a different job from chloroplasts. But even when accepting Drexler's statement at face value, there still seems to be a problem with solar cell collection at our present level of expertise. Elkington points out that in manufacturing and deploying solar energy devices, you often end up using more energy and sterilizing more ground than makes economic sense.[18]

Compare this single-track energy capture with the multifunctional character of photosynthesis, and it is clear that we have a long way to go, in attitudes if not in technology.

> Although the astonishing insights and techniques of present-day science have fairly well analyzed what takes place in a green leaf, man can make sugar glucose in a test tube only by slow stages, laboriously, and with temperatures and chemicals which would quickly destroy a living cell. And the result is only a few insignificant grains of sugar. While the scientist labors for days with his precarious experiment, green leaves outside his window, without the slightest fuss, are making an abundance of sugar within thirty seconds of a flash of light, and at a balmy temperature.[19]

Even as they make this sugar for their own sustenance, trees are performing a multitude of other functions for the environment as a whole. Forests are vital watersheds that absorb excess moisture, renewing underground water supplies and preventing floods. They anchor topsoil. They absorb carbon-dioxide and release oxygen, temper the heat of the sun's rays, and send their roots deep into soil and rock to bring up essential minerals. Their bark and falling leaves create a wide range of habitat for microscopic life forms, vital links in the food chain of other species. Their fruit and leaves provide food directly to animals and man, while even dead trees provide shelter to sprouting seeds, insects, fungi and bacterial life. And all this is done while collecting free energy from the sun. Writing in 1965 Platt was able to tell us that all the oxygen in the vastness of our atmosphere is filtered through chlorophyll every two thousand years, and that all the carbon dioxide in the air passes through green leaves every 300 years.

> This spells a power operation by chlorophyll that generates a hundred times more kilowatts of energy a year than all man's hydroelectric plants, coal, gas, fuel oil, and atomic-energy generators combined.[20]

The wholesale destruction of trees by man in the last few decades has made a frightening dent in these beneficent services, and there are many who feel that the only solution to the greenhouse effect is to plant untold millions of trees. In October 1988, an American power company gave substance to this conviction by announcing its contribution to the cost of planting fifty-two million trees in an agroforestry project in Guatemala.[21] Many other such schemes are under way, witness to the growing realization that trees are essential to the continuance of life on the planet.

But the point I am trying to establish here is the interesting multiplicity of function displayed by trees in the capture and storage of energy. The mammoth power-stations that produce so much of the air pollution in industrialised countries have only one function - to generate electricity. The steam that billows out of cooling towers, for example, heats up the atmosphere at huge expense without benefit to anyone. I believe that in addition to establishing a multitude of smaller, local, and more varied methods of producing energy, we need to find ways of making energy sources do a number of reciprocally useful things at the same time.

An interesting example of this is the experimental 'Ark' concept, developed by the New Alchemists under the leadership of John Todd in the late 1970s. In its early stages it took the form of a 'bioshelter' consisting of an 18-foot geodesic dome containing a fish tank, with a few salad plants growing around the edge of the pool. Later projects were more ambitious, and though self-sufficiency in food was never a prime objective in the experiments, considerable quantities of fish and vegetables were in fact produced. Two such Arks were developed, one on Cape Cod, Massachusetts, and a second on Prince Edward Island in the Gulf of St Lawrence. Both were defined by the New Alchemists as 'autonomous structures, in that they have had no reliance on an outside energy source like fossil fuels, but regulate their own internal climates through the collection and storage of solar energy.[22] The Cape Cod Ark is now used for food production in all seasons, while the Prince Edward Ark also contains living quarters.

The architecture of the Arks was explicitly solar, both buildings facing south and presenting an upper wall of flat-plate collectors to the sun.'But the New Alchemists took things a step further by incorporating a fish farm.'

The first stage in the Prince Edward Island fish culture cycle starts with two rows of forty translucent tanks - which contain dense 'blooms' of algae. The algae provide food for the fish, but they also act as an efficient solar-energy collection surface, while the tanks serve as heat-storage units.[23]

Prince Edward Island is an exposed headland overlooking the St Lawrence, with winter temperatures that freeze the river each winter. Elkington describes how Todd and his team took the daring step of establishing their Ark in this inhospitable terrain because, if the concept worked there it would work anywhere. When the temperature drops in the Ark, the fish tanks re-radiate the stored heat into the greenhouse area and thence into the living space behind the greenhouse. When the sunlight fails, an advanced Hydrowind windmill is used to provide electricity for heating and for the laboratory sensors and other equipment. Elkington notes that the fish tanks have achieved unprecedented levels of biological productivity, yielding the highest recorded fish production per unit volume of water for a standing water body.[24]

So what is it like living in a twentieth-century Ark? It is, according to Todd, 'an extraordinary experience. Living and working in a structure where the sun, the wind, architecture and ecosystems are operating in concert has affected most of us. It seems to foretell what the future could bring. Each of us involved in the project would like to one day live in an Ark-inspired bioshelter ... ' The Ark's inhabitants, Todd suggested, had come to feel a sense of wholeness, of ecological integrity. 'Working with vegetable crops, flowers or trees, or tending the fish while a howling blizzard rages outside, is exciting. Watching the sensors in the laboratory read out the Ark's various functions and monitor its health while the cold wind blows, connects one consciously to the forces sustaining all life.'[25]

While the thought of living in such an ark is not universally appealing, and while at this stage the arks are beyond the ability of the average person to operate and maintain, the concepts embodied in bioshelters are sound enough. Simpler developments may follow. One remembers that solar architecture had been brought to a fine pitch by the ancient Greeks, that Swiss peasants benefited from the body-warmth of their cattle by letting them winter in quarters adjoining their own, and that the buildings of many primitive peoples were far superior to our own in insulation and the storage of heat. We have much to learn from the past.

The current practice among city dwellers of growing plants in every imaginable corner of their homes suggests that mankind is unhappy in an environment too far divorced from natural things. Architects and builders are daily becoming more sensitive to the contribution which solar energy can make to their energy budgets, and perhaps architectural designs of the future, by borrowing some of the ideas of the bioshelter, will make life happier while reducing the profligate expenditure of energy of our major cities. Meanwhile there is much experimentation to be done, and once again, diversity is of the essence. Climatic and geographical norms will decide in the end which type or types of energy production can best be enmeshed in any given area.

f.) Solar Power Options

The main energy-capturing technologies at present in use can be thought of in two main groups:

1. Those which harness direct flows of solar energy in the form of either heat or of photons. Heat collection can be either 'passive', such as in simple roof panels for heating water, or 'active', such as the high-tech hybrid systems which concentrate sunlight to produce both heat and electricity. Photon capture technologies may take the form of either photovoltaic cells or photochemical cells. The latter use photons to drive chemical reactions.

2. Those which tap into indirect flows of energy emanating from the sun. These are hydropower, wind, and wave energy.

Elkington suggests that the most significant solar technology in use at the moment involves the exploitation of the energy stored in *biomass*. 'This generally implies extracting stored solar energy, by whatever means, from plant materials or from micro-organisms.'[26] This is interesting, for as long ago as 1976 Brian Ford spoke of 'Microbe Power' as 'Tomorrow's Revolution':

> Every living cell is a controlled powerhouse of energy. ... each one produces energy from the breakdown of food materials. The result is a rate of production no man-made factory could emulate, and it could provide us with an unlimited prospect for biological technology. ... The lesson we should learn from this is twofold:
>
> 1. It is wasteful to apply external energy to any reaction that is capable of fuelling itself.
>
> 2. It is wasteful to utilize industrial processes for any reaction that microbes can carry out.

The exothermic, energy-giving reactions of life are capable of being harnessed for all manner of industrial tasks and it is now, with the demands for energy approaching extreme limits, that we should begin to investigate them.[27]

Microbes, Ford insisted, are well adapted to undertaking startlingly efficient processes, and merely need to be recognized and channelled into new areas to become important sources of power for mankind. 'For a great many industrial processes the use of endogenous energy, already locked away inside the chemical structure of the raw materials, makes external demands on oil or other fuels unnecessary.'[28]

Denis Hayes, Director in 1980 of the US Solar Energy Research Institute, including in his definition of solar resources all the above options, was convinced that by the year 2000, renewable-energy sources could supply some 40 per cent of the global energy budget, with a figure of 75 percent on the cards for 2025.[29] The operative word is 'could', for sadly this scenario is dependent to so large an extent on economic power-play as to make any kind of forecast impossible. However, the options are undoubtedly there, do we choose to develop them. Renewable energy sources include what is generally known as 'hard'as well as 'soft' technologies, and, contrary to general belief, are not necessarily without hazard.

g.) Energy from the Oceans

Tides

Using tides to generate electricity is achieved by building a dam across a narrow arm of the sea where the incoming tide raises the water level by more than five metres, creating an artificial resevoir. As the tide ebbs the water behind the dam is released through pipes to turn a turbine. The snag about this system is the 50 minute variation of the tides in the 24-hour cycle, and that the ebb flow will not always coincide with the demand for electricity. The consequent breaks in production necessitate large-scale storage of electricity.

The first large-scale commercial plant was established in 1967 at the mouth of the river Rance, in Brittany, but interest continues to grow, and over 40 possible sites have been identified. While the newly-formed lakes may offer attractive tourist possibilities, there are disadvantages even to this apparently innocuous procedure. Gourlay points out that the mean water level behind the barrage is now higher than previously and the waters of the basin are exchanged less frequently with the sea.

In consequence, sewage and industrial waste formerly washed out to sea tends to linger and, unless given additional treatment, can rapidly cause pollution. Heavy metals in particular are caught up by suspended sediment and collect on the bottom, where movement is also reduced, thus affecting bottom-living creatures. Add to this changes in the salinity of the water and it is obvious that, whatever precautions are taken, some effect on fisheries is inevitable.[30]

Wave movement

Gourlay mentions three proposed types of energy converters using the huge power of waves:

1. those using the vertical rise and fall of waves to force air or water into operating turbines;

2. those using the rolling, pitching or heaving action of waves to activate cams which work similarly;

3. wave focusing, which concentrates available energy at one point. One type of this, 'surge funnelling', channels shoaling waves into a storage basin and the hydraulic head produced is used to run a turbine. A further development, proposed for Mauritius, heightens a coral reef artificially to create a hydraulic head within the reef. This can then be used to drive a turbine at the outlet. This device, known as 'passive'wave energy, embodies extreme potential hazards, since changes in the natural working of marine life could affect the whole ecosystem, including damage to fisheries and the corals themselves.

Wave power has the advantage over tides, of greater persistence and no unavoidable breaks in supply. Even bad weather can be turned to good use as waves are conveniently more plentiful in winter when demand for electricity increases. Scientists have estimated that a quarter of the electricity used in the UK could be obtained in this way without extra energy storage and that, if the wave power around the coasts of Britain and Ireland was fully used, it could supply more than five times the present UK demand.[31]

This all sounds ideal, but once again there are snags. Large-scale harnessing of wave energy could produce a calm sea between the point of extraction and the shore. While this would make navigation easier and decrease land-erosion in the immediate area, coastal sediment previously carried away by the sea would accumulate. This, in conjunction with the wave-breaking machinery itself, could result in changing currents that caused erosion elsewhere. Once again we are faced with the incalculable and sometimes frightening interplay of forces over which we have little control, and whose long-term effects we do not really know. What is interesting, here, is how closely a technology can mirror the geographical characteristics of a region. Great Britain, for instance, with its deeply indented coastline and relatively small land mass, would seem to be ideally suited to the generation of electricity through the action of waves.

Sea Thermal Power

A high-tech large-scale energy source, known as OTEC, (Ocean Thermal Energy Conversion), uses the difference in temperature between hot surface waters and cold waters in the depths as a means of producing energy. The technical problems are considerable, as the structure has to be moored in deep water, and requires powerful pumps to supply both types of water. Cables must then be laid to transmit the energy to land.

This type of electricity generation would seem at first sight to fulfil the systemic requirement of multipurpose usage, since the cold water of the deep ocean is rich in nutrients that could be used for setting up fish farms at sea. On the industrial level, by extracting hydrogen from sea-water by electrolysis and combining it with nitrogen from the air, the plant could produce ammonia both for its own operation or as a saleable commodity, while yet further developments would be the production of alternative fuels such as methanol and of fresh water by desalination. Unfortunately the dangers are such as to presage ecological disaster.[32] The implications of Ocean Thermal Energy Conversion - too complex to be discussed here - are such that even under 'normal working conditions' large-scale disruptive effects on marine life,and a gradual build-up of toxicity are assured, while 'accidental catastrophic incidents'cannot be fully excluded. Gourlay compares these in magnitude to Chernobyl and Bhopal. We've seen enough of very-large-scale enterprises in the production of energy to realize the magnitude of disaster when some small thing goes wrong in a mammoth plant. In energy, then, as in industry and farming, multiplicity and diversity are key elements in keeping both pollution and destruction under control. They are also key elements in the systemic approach.

Fuelwood

Some 2,000 million people still rely on firewood for their fuel, a source of energy that is rapidly assuming crisis proportions. The roots of the fuelwood crisis are complex, ranging from population pressure to the clearing of forests in pursuit of projects that will earn foreign exchange. These include commercial plantations, large-scale destruction of forests to make way for cattle, and the production of cash crops.

In terms of adverse ecological impact, few things are more serious. As fuelwood supplies become exhausted, animal dung is burned instead of being returned to the soil. Fertility is reduced, forcing the clearing of more and more forest for cultivation. Areas once protected by woodland are left exposed to wind and flood; erosion and desertification follow. Carbon dioxide levels mount. Climatic patterns are disrupted, and the 'greenhouse effect' begins to make itself felt.

Encouraged by the World Bank, several governments have started 'social forestry' programmes, ostensibly to provide villagers with firewood trees. Unfortunately, the bulk of the trees have almost universally gone for commercial uses, such as rayon production and paper milling.[33]

More heartening is the report that successful reforestation schemes have been implemented by local groups, acting at the village level, and often working without outside funding. Such groups include the Chipko movement in India, and KENGO, the Kenya Energy Non-Government Organization.

It is probably true to say that no activity in which any and every citizen could be engaged is more important than the planting and care of trees, in any and every place where suitable species can be coaxed to grow, and the wide diversity of species allows for cultivation

in most parts of the world, however inhospitable. Goldsmith and Hildyard express the concerted view of many experts that 'the only real solution to the fuelwood crisis lies in putting an end to further deforestation, and in implementing massive village reforestation programmes, using non-commercial species.'[34]

PART TEN

The Information Society

CHAPTER FORTY NINE

10.1 Principles

Summary: The Communication Sciences give us the capacity to build society in accordance with living principles if we have the vision and moral fibre to use them correctly instead of destroying ourselves.

The Information Society

And so we come to the final and crucial section of this investigation: How can we best devise an immediate and effective nervous system for the planet, using available technology? And have we the moral fibre to overcome the enormous political, social and economic constraints involved?

I hope I have been able to show that it is both possible and desirable to progress towards living principles in the organization of human society. I believe we have already taken some of the transitional steps towards such organization, and that full transition is essential if we want to survive. So we have to ponder deeply on two vital issues:

- Do we care enough what happens to posterity to work for it?
- Is there any means of speeding up the process to avoid catastrophe?

For the first we have to rely on the innate goodness of many millions of humankind, on the embryonic urge to survival of the infant Gaia, and on the divine sub-structure of creation itself.

For the second, the answer lies in the integrative capacity of the information sciences, applied with all the will and wisdom at our command.

It has of course been argued very convincingly that man's salvation lies in space; that resources are limitless; that space holds matter, energy, and room enough for projects of vast size, including vast space settlements; that computer-controlled molecular machinery will make possible the building of self-replicating assemblers, enabling us to produce anything and everything we need from the unlimited resources of space.

> Replicator-based systems will be able to construct worlds of continental scale, ... made of strong, carbon-based materials. With these materials and water from the ice moons of the outer solar system, we will be able to create not only lands in space, but whole seas, wider and deeper than the Mediterranean. Constructed with energy and materials from space, these broad new lands and seas will cost Earth and its people almost nothing in terms of resources. The chief requirement will be programming the first replicator, but AI systems will help with that. The greatest problem will be deciding what we want.[1]

And this of course is the crux of the matter: What, really, *do* we want? It is true that a way to by-pass terrestrial limits to growth is extremely heartening. Drexler remarks that 'Establishing an era of universal wealth will require only that the vast, unclaimed resources of space someday be divided in a way that gives everyone a significant share'.[2] But does our handling of resources on planet Earth give reason to believe we shall do any better elsewhere? I think not. Wherever we go we take ourselves with us, and our ethical record is far from encouraging. But do we really have to think in terms of space before we can make our own Earth operable? Is this what we want? My own dream is much more modest. Not some high-tech future in space, though that may come, but to help make our own beautiful planet work, here and now, in the lifetime of people alive today. We do not have time on our side. Lewis Thomas once remarked that what man needs over and above anything else is a future. It is this future I am working for - perhaps the next twenty years. After that, if we are still here, and all these wonderful things can come to pass, the sky is the limit indeed. It would be nice to know, as we sailed unfettered with the solar wind, that we left here at home a beautiful, functional Earth whose ethical integrity glowed like a sun in the universe.

Institutions designed to further free speech, freedom of worship, due process, and the rule of law have all helped in the past to bolster human frailty. But change comes upon us too fast. Institutions designed to maintain a stable world are breaking down. We have to find a way to embody their principles more fluidly. And we will. We need a simple, abiding belief that it is possible.

Drexler observes that we have succeeded in technical matters because we have learned to build institutions that draw many people together to generate and test new ideas.

> 'Erratic people of limited competence can join to form stable, competent institutions. ... No one [person] knew how to get to the Moon, and no one ever learned, yet a dozen people have walked its surface.'[3]

The same holds true, I think, for social concerns. Just how many of our current institutions will, or should, survive is still unclear. But the principles remain.

CHAPTER FIFTY

10.2 The Ethical Road

Summary: Each one of us is responsible to all the others, for everything.

The Ethical Road

Solutions to today's problems offer such a wealth of ideas, proposals, redundancies and counter-proposals that only complete openness to criticism, discussion or support can achieve a workable balance, sorting the grain from the chaff. And here again, conventional means of correlating the flood of data threatens to overwhelm us. Truth is of the essence - communication the means. And fortunately electronic research laboratories are building ethical principles into their work from the very start. Almost any serious reading endorses this.

Typical is the work of Jerome B. Wiesner, past President of MIT, of Nicholas Negroponte and his colleagues at MIT's Media Lab, of Japan's Koji Kobayashi, and of countless others in the field. All are overwhelmingly in agreement on the need for an ethically healthy communication system. That the business world does not always see it that way is in no way the fault of the researchers. For them computers are for people, all people. The information sciences are to be made so user-friendly that individual people everywhere can handle them, no matter what their degree of literacy or sophistication. And if this assertion raises an eye-brow or two, then share with me Mr Drexler's delightful words:

> Since ancient times, villagers have used the molecular machinery of yeast, seeds, and goats without molecular-level understanding. If such complex and unruly things as goats suit primitive ways of life, then other forms of molecular machinery will surely qualify.[1]

So research is aimed at machines responsive to personal needs and adaptive to personal circumstance. These machines are highly selective - one might almost say 'discerning' - designed to select precise and user-appropriate data from the overwhelming masses of information pouring down on us. Above all they are to be universally interactive, putting anyone in touch with anyone anywhere in the world. And the more advanced the technology, the easier of attainment this goal becomes.

Personalization is at the centre of this research. The many technologies are treated as a single craft. The subject is a neuron-like interconnection of human beings.

To this end much of the initial work has already been done. As long ago as 1985 Japan's Koji Kobayashi envisaged in great detail how he expected innovations in Computers and Communication (C&C) to affect society's enterprises, suggesting structural networks ranging from the smallest units in the home and office to global networks geared to the interaction of people all over the world. His book *Computers and Communications* (English translation published in 1985 by the MIT Press) presents clearly what is meant in Japan by 'the communication society', a society in which the integrating power of modern information technology is made available to the entire society. Kobayashi acknowledges freely that much of the technology he describes has come piecemeal from other parts of the world, and especially from America, but his especial genius seems to lie in a vision of its potential, and of its practical integration in a single, coherent global system.

In the words of MIT's Jerome B. Wiesner:

> The vision that underlies Dr Kobayashi's account is one of great power and hope. It articulates a future in which individuals everywhere can communicate with each other, in which knowledge is universally available, and C&C information systems overcome not only barriers of distance but barriers of language, leading inevitably to a single, coherent, and competitive but friendly world.[2]

It is only one step further in the same process to accomplish the friendly, mutually helpful world of my own imagining. But language alone, while certainly a tough barrier to understanding, is only the surface manifestation of deep cultural values. All social systems, from the so-called 'Western' to those of the Soviet Union or the far East, refer their value judgments to their own very different value systems. Because of these underlying differences, even the interpretation of words differs in each case, and the attitudes which they invoke may be quite different from what the speaker intended - which is one of the reasons why so many international conferences break down. There is reason to believe that future computers will progress to a point where they can assist us in overcoming even these difficulties, but from where we are now it would seem that an overwhelming impulse toward self-preservation must be our most powerful tool, and that this has at last begun to energize our understanding, one for another. The biosphere recognizes no divisions into ideologies, power-blocs or nations. Faced with the threat of imminent global catastrophe, dividing lines are beginning to dissolve, politicians to recognize that ecological cooperation with the enemy is the only means to survival, and military might to lose its credibility at centre stage. Integration of all his resources in the manner of a living system has become the only choice now open to man. The electronic revolution has made it possible. The process has already begun. And the speed of its evolution catches the breath.

Recently I passed a gateway where, in the garden beyond, someone was restoring a two-wheeled horse-drawn 'fly' with loving care. Instantly I was back over three-score years and ten, to a childhood when people drove about in just such fragile craft. Motor cars were still unknown in our part of the world. So were telephones, while radio and television belonged to a future that was not yet part of our imagining. All the way from pony-trap to computer power in the span of one short life!

Lounging at ease in my son's superb BMW, more than seventy years later, the memory of a childhood summer's day came back: an innocent drive in the sunshine; an angry ostrich, perhaps defending his nest; a startled pony; flying wheels of the trap; exhilaration; a little prickle of fear. Urged to a gallop by our anxious nurse, our pony was doing his sturdy best, while my sister and I crouched low in the seat, ducking from the beak of the furious, striding bird.

Today, as I write of fusion, of a world becoming one world, of everyone's dependence on everyone else for the very basics of life, I marvel how insulated we children were, so short a time ago, and how far mankind has travelled in so short a time. Last night's programme on television, incorporating music by the London Chamber Orchestra, showed today's children lighting candles all over the world in a drive for peace. We learned that since the end of the Second World War in 1945, twenty million people had been killed in ongoing conflicts. Sixty million had lost their homes and been rendered destitute. Families had been torn apart. Children unwillingly abandoned by their dead. And Peter Ustinov quoted Dostoevsky's prophetic words, carved over the portals of the Red Cross Headquarters in Geneva:

Each one of us is responsible to all the others, for everything.

We have a long way to go to translate this responsibility into action. But the means are now there to our hand. Rightly used, with the will of the world behind their power, communication sciences can open the way. The input of each can at last be weighed and assessed.

CHAPTER FIFTY ONE

10.3 Earthvote

Summary: How each man's responsibility to all the others can be translated into the patterns and structures of society.

Earthvote

Because humankind is part of a larger living system and this system is dependent on its every individual unit for healthy functioning, every man automatically carries responsibility for the Earth, which is in this sense only a wider aspect of himself, whether he is aware of it or not, and whether he accepts such responsibility or not. He himself is at once a synergetic assembly of smaller functioning entities, a man, and a small entity in the much larger synergetic entity which is the Earth. There are no boundaries. But in order to realize his full potential as a man, he must translate this unwritten responsibility into the patterns and structures of human society. It follows that if he has a right or a duty to vote in his own interest on any aspect of social organization, he also has both a right and a duty to vote in the interest of the Earth, on each and every conceivable issue. And this vote, which I shall call the Earthvote, should be used in conjunction with his own in any action or debate in which he is involved. I suggest that we now have both the knowledge and the computer-power which will enable him to do this. And I suggest that all societies should, by common consent, so arrange their legal and social norms as to incorporate each man's Earthvote automatically in decision-making at every level.

How would we do this?

Supposing, in addition to his computerized voting power as an individual, each man carried a programmed Earthvote, so that whatever discussion was under way, voting would automatically incorporate ecological priorities - what was, in fact, 'good' for planet Earth. In some cases what was 'good' might be debatable, but we'll come back to that.

Let us suppose, for instance, that there were a strong lobby in parliament in favour of a lucrative, but highly polluting, health-endangering industrial project, in an ecologically sensitive area. Let us suppose, further, that a government lobby were in favour of massive subsidies for the project, because of its promise in providing jobs, and because of the substantial tax revenue to be expected from it. There would be strong arguments in favour of an enterprise that could be expected to give a boost to the economy at a time when many people were starving, and the government would be under strong pressure to intervene. There would perhaps be equally strong arguments concerning adverse effects on the health of workers and their families, on local supplies of drinking water, on neighbouring small holdings, and on the environment as a whole. How would the debate be settled?

We'll keep the numbers down to simplify matters. Let's suppose six people were involved in the debate. Then a vote of, say, 5:1 in favour of the project would read 5:7 against it when each man's Earthvote had been incorporated and the needs of a sustainable planet had been taken into account. The Earthvote of all the voters would have outweighed the majority vote of the 5 originally in favour of the project.

Supposing the vote had been 3:3. Then the Earthvote would result in a count of 3 for the project and 9 against. And the decision would be in favour of a sustainable society, which is what we need.

If all six had voted for the project, i.e. 6:0, the Earthvote would need 1 extra vote to block it. So we would need to allow an Earthvote-plus-1 factor in the programming of all our computers.

If, on the other hand, an integrated and unbiased review of all the relevant facts revealed that the industry was essential to the development of, say, solar power, would contribute thereby to a local solution of the energy crisis, could be non-polluting if certain simple procedures were followed, would then offer healthful activity in the common interest to large numbers of local people, and could be so managed as to inflict minimal damage to the environment itself, the Earthvote loading might give a different result. Not the most vociferous protests of the local crank could then affect the issue.

Of course while the profit motive remained the top priority in human affairs, such an arrangement would be unlikely to find favour, but the urgency of positive action in the interest of human survival is rapidly becoming known, and once personal fear enters an equation, priorities can change overnight. So the point of the exercise would not be to control the activities of large concerns hell-bent on a profitable enterprise, nor to confound economists struggling with problems of the national budget. Instead it would essentially be designed to assist local bodies in making wise decisions on activities in their area, even when they had insufficient personal knowledge to do this.

We've been talking about a global network giving everyone access to the sum of human knowledge. This network could be so programmed that Earthvote-loading was incorporated in it at all times, and though our knowledge of ecosystem operation is still fragmented and incomplete, we would at least be basing our actions on the best that was known at any given time. Moreover the Earthvote decision would mirror as closely as human knowledge allowed, the subtle and changing patterns and needs of the whole global system. The new multiple-network computers discussed above, based as nearly as possible on some of the processes of the human brain, and able to combine both bottom-up and digital procedures in their operation, should prove ideal for this kind of work.

Having access at any and every moment to the most up-to-date knowledge from all parts of the globe, and having the capacity to integrate and assess vast masses of relevant data in an instant of time, the network would be in an infinitely better position to influence decisions in the interests of the whole organism than any less informed body could be. In addition, computers could be programmed to flash on the screen the pertinent information on which the Earth-vote was predicated, explaining why certain procedures were more beneficial than others, and what the global consequences were likely to be in the event of wrong decisions being taken. Sensible local bodies would of course use this facility at the beginning of any debate, to enable them to vote wisely in the light of the best available knowledge. Faced with clear evidence that his own best interest depended on the wellbeing of Earth's life-support systems, even the most determined self-seeker would be likely to go along with the right decisions. Where greed or stupidity overruled even the clearest guidelines, the Earthvote *recognized by society as a whole* would ensure that no individual or vested interest could rock the boat.

The network could also do a quick scan of similar or near-similar conditions elsewhere on the globe, and the methods and solutions that had been employed there, and could flash these onto the screen as possible alternatives to the project under debate. Any person or group of people, from the small town cooperative to city councils, corporations or governments, would thus be stimulated to creative thinking by experience wider than their own.

And the beauty of it all would be that decisions would not be imposed by any outside authority; nor would the voting power finally reside in any party or group. It would reside in every individual's cohesive membership of the same living body, thus fulfilling the same physiologically beneficial function for the Earth as cells do in living organisms. It would embody the intrinsic right and duty of every human being to contribute his best potential to the whole.

One of the most difficult problems in the history of human society has been and is, how to reconcile individual freedom with the common good. The automatic use of each man's own intrinsic Earthvote overcomes this difficulty. Simply by being himself, every man becomes his own limiting factor in decisions affecting his basic wellbeing.

This, as it seems to me, is the beginning of true self-government in human society, in a way very like that in which organisms operate. We saw early on in this book, in the work of Jacques Monod, that even at the level of a protein molecule, microscopic living systems have the capacity to 'choose' a reaction, and that the consistent criteria for such a choice is their contribution to the coherence of the system.

It would be strange indeed if human beings proved incapable of a similar wisdom.

CHAPTER FIFTY TWO

10.4 Computers and Communication (C&C)

Summary: The ever-increasing range of electronic devices coming on the market, each translatable into the others, generates so pervasive a network that, subject to the limits of financial constraint and political expediency, no one need be isolated anywhere in the world.

Computers and Communication (C&C)

Anyone with a basic knowledge of modern communication systems understands just how inter-related the electronic medium has become. Messages travel on the ground, in the air, under the sea or by satellite. They link factory, office, home, motor car or aeroplane, using every electronic device so far evolved, including phone line, coaxial cable, fibro-optic cable, microwave, satellite, air broadcast, or physical storage medium such as tape or disk, the system made interactive at every point by the integrating power of the computer.

Not only is an ever-increasing range of electronic devices coming on the market, but by the digitalization of all these media they have become translatable into each other. The ubiquitous microchip ensures that computer bits are transferred from one medium to another without loss or fuss, so that any message, sound, or image may be edited from anything into anything else. And the inevitable consequence of this interaction is to generate new combinations and technologies beyond anyone's power to predict or legislate. Already remote imaging from space by civilian services is approaching military standards (making a mockery of trans-border legislation) while fibro-optic cable makes the band-width bottle-neck a thing of the past. Subject only to financial constraint and political expediency, every home could already acquire its own broadcast capability. Every hut in Africa could have its terminal, the body of human thought within its reach. And every hut in Africa is beginning to want just that, without the slightest conception of what it means.

The seeds of this revolution were cast years ago. It is fair to say that without the advent of radio, the powerful National Consciousness movements of underdeveloped nations would never have got off the ground. But band-widths on the radio spectrum were limited, with the result that favourable allocation of the available space inevitably went to the technical and industrial elite able to use them. Benefits thus accrued consistently to those in whose hands advanced knowledge meant power and wealth, resulting in serious disadvantage to Third World countries, as well as to the poor of developed communities everywhere.

This situation is changing fast. History suggests that, given the opportunity, man will always exploit his fellow man, but the nature of the developing technologies forbids such practices. So pervasive are the new media that 'owning' information becomes as impossible as trying to keep it secret, giving rise to considerable alarm, and a wealth of legal confusion no-one can even begin to resolve. The chaos predicted for the future is already here. It is no longer tomorrow. It is now. The poor demand a share of what they see, and by their numbers begin to embarrass the rich. Leaders sometimes profit by this to gain control of wealth and power, and only a half-wit could be sanguine about it. But that the poor themselves should have a say in controlling their lives can no longer be held in doubt. By the sheer weight of numbers starving people become a political force.

By the same token, the new media strip away the camouflage from some of our hypocraces. When, in famine-stricken nations such as Ethiopia, grant-aid is channelled into military might instead of into food, their leaders are no less worthy of condemnation than are those who profit from selling them weapons. Before the ubiquity of the new technologies, president and pauper stand revealed alike.

But there's a positive side to all this. One half of the world now sees how the other half lives. For the first time the 'haves' really 'see' the plight of the poor in a meaningful way. Starvation invades their homes at the touch of a television button. Disease and death on their own door-step become each day's visual and empathetic experience. Statistics transform themselves into people, and situations such as those in Ethiopia, Iraq, Bangladesh and most recently in Ruwanda, draw a public response beyond all expectation. The death of anyone's child is suddenly everyone's concern. In Bangladesh in April and May of 1991, death was counted in hundreds of thousands of anguished people, but when a dolphin saves one drowning baby on the far-off beaches of Bangladesh, the world sighs with happiness.

Given real understanding, then, human compassion knows no bounds. We are forced to the conclusion that for the vast majority of human-kind, all that lies between a man and his compassion is survival in the economic rat-race, the scramble for ever more money, and the universal belief that there is no other way. We begin to glimpse alternatives. Slowly we are moving towards them, gathering momentum as we go.

CHAPTER FIFTY THREE

10.5 The Connectionists

Summary: Recent computer research includes the development of a number of radically new models, based on neuron-like units linked together in multi-layered interconnected nets.

These models are of particular interest in the context of an emergent global consciousness.

The Connectionists

For anyone interested in an emergent global consciousness, one of the most interesting developments in the computer field is the work of a group of scientists who call themselves the 'Connectionists'. They come from a wide range of converging disciplines which includes, among others, computer scientists, neurophysiologists, physicists, psychologists, linguists, mathmeticians and philosophers. All these men are interested, each from a different angle, in the working of the human brain and in the emergence of mind from a vast array of interconnected neurons. Their work entails the development of a radically new type of computer model, based on neuron-like units *linked together in multilayered interconnected nets*. This approach had its origins in work that began as long ago as 1950, but at that time limited results caused a temporary shift to research programmes based on words and symbols. The recent linking of units in *inter-connected nets* solve many of these earlier problems.

Multi-layered nets embodying 'hidden units'enable these models to make representations of the outside world, thereby increasing exponentially the complexity of tasks they are able to perform. The scope of research is wide open, with scientists concentrating on that branch of study that bears most directly on their own work. It includes the search for better understanding of how a brain thinks, the building of AI machines on neural principles, and the application of these principles in the study of human society. Eventually, says Drexler, 'neurobiologists will use virus-sized molecular machines to study the structure and function of the brain, cell by cell and molecule by molecule where need be ... [but] neural simulation can succeed without such insights.'[1] Current research by the Connectionists would seem to suggest that the new neural nets are doing just that.

As I see it, this development could hardly be more important, because if these men are successful - and I haven't the slightest doubt that they will be - we shall have at our disposal machines capable of processing enormous loads of complex social input in a manner that has at least some characteristics of the human brain, but is immeasurably faster. The potential benefits to social integration are immense.

One of the ways in which a neuron-type network differs from a conventional digital computer is that it does not require a central processor. While the latter processes data sequentially, a few bits at a time, but very fast, the neuron-like units in the experimental nets act on data all at once, as do the neurons in our brains, bringing the entire system to bear on the task. Each neuron takes in signals from the other neuron-like components, adds them up, and decides on the basis of the answer whether to send out a signal of its own. William F. Allman suggests that the neural units in the machines are analogous to people sitting on a jury talking among themselves, trying to influence each other to decide one way or another:

> When an input comes into a neural network, there is a mad jumble of changing votes and opinions at first, but eventually all the neurons settle on a decision and the machine produces an answer.[2]

Thus the state of any one 'neuron'depends on the states of all the other neurons connected to it, which are in turn reciprocally dependent on input from the interconnected units of the whole system. Analogy with human society is irresistible.

It has been shown mathematically that a network of artificial interacting neurons could be thought of as a physical system whose energy is decreasing. Over time, each system as a whole evolves from a high-energy state to a low-energy state.

> At the point of receiving an input, the network is in a high-energy state. As the neurons talk among themselves and produce a decision, the system evolves toward a lower-energy state. Hopfield's neural-net 'thinks', in other words, not in the logical, step-by-step manner of a typical computer, but more like a New England town meeting, where an initially heated debate gradually settles into a decision.[3]

This concept is of particular significance to me, because, in my unskilled layman's imagining, this is just how a global network will need to operate. While the popular notion of an enormous centralized global computer of the top-down variety may still be valid for those statistical tasks involving logical reasoning and arithmetic, even for these tasks decentralization seems preferable. A system composed of many machines, with elements designed and programmed by many millions of people, is likely to be resistant to control or abuse by any single group. Once again diffusion and diversity are the essence of healthy development.

This does not preclude the use of top-down and bottom-up principles working in concert. We know that this happens in the brain, and the same will surely prove true for computers, but for the social integration of millions of diverse human beings, the connectionist model would seem to offer unparalleled possibilities.

Neuroscientists have some reservations about these new models of the mind, because they are of necessity grossly over-simplified. To begin with the interplay of neurons in the brain involves chemical as well as electrical components, and each individual neuron is in itself a complex living cell containing *billions* of large interacting molecules of thousands of different kinds. While the structure and function of many of these molecules is now understood, that of others remains unknown. What is clear is that a neuron is emphatically *not* just an assembly of switches. Moreover each of the brain's hundred billion neurons may be connected to as many as a hundred thousand others, and the synapses rearrange themselves continuously in response to incoming stimuli. No-one yet understands fully just how this comes about. But already the new models are producing sufficiently impressive results to intrigue an ever-wider body of men.

We know that very large numbers of elements of any kind give rise at a certain threshold to complex patterns of behaviour familiar to physicists as 'emergent properties', and that it may in part be this dynamic interaction of very large numbers of neurons that gives rise to both consciousness and thought. 'Though they are the consequences of millions of microscopic relationships, emergent properties often seem to take on a life of their own', says John Hopfield of the California Institute of Technology.

> Suppose you put two molecules of gas in a box. They move around the box, and every once in a while they collide. If we put 10 or even 1,000 more molecules in the box, all we get is more collisions. But if we put a billion billion molecules in the box, suddenly there's a new phenomenon - sound-waves. Nothing in the behaviour of two molecules in the box, or ten or 1,000 molecules, would suggest to you that a billion billion molecules would be able to produce *sound waves*. Sound waves are a collective phenomenon of a complex system.[4]

We still do not understand just how this works, either for the molecule or in the brain, and the manner in which the interplay of billions of neurons gives rise to mind is still a profound mystery. What we do know is none the less sufficient to inspire some very interesting connectionist models, able to perform complex tasks of learning and association that are both beyond the scope of the conventional digital computer, and different in kind. Researchers have developed multilayered nets that can draw inferences, make generalizations, form categories or even organize themselves into groups specialized for different kinds of information processing. This last is crucial, for it involves the ability of the nets to rearrange the connections between their neurons in response to increased input, increasing the number of pathways they use to process incoming information - a function characteristic of neuronal activity in the brain itself.

The evidence suggests that it would be rash to infer a neural system's functioning as a whole from the behaviour of individual neurons. Commenting on the work of brain researchers such as Sejnowski and Lynch, Allman observes that:

> as more and more details are discovered about how our brains process sensory inputs, store information, and retrieve memories, scientists will have to shift away from studying how neurons act individually toward studying how groups of neurons interact as an integrated system.[5]

This, says Allman, is a change that scientists from many other disciplines are beginning to make, and neuroscientists are realizing that as in the weather, the nation's economy, and other extremely complex systems governed by interactions among many simple elements, the activity of the brain as a whole may be quite different from the sum of its individual parts.

The direction in which these studies seems to be moving can only be a healthy one, for in every case they attempt to mirror neuro-biological processes, evolved, adapted and tested over millions of years. In the present turbulent century, and more particularly in the present decade, the behaviour of human beings has resembled nothing so much as the mindless molecules-in-a-box that Hopfield describes. The connectionist research may be expected to throw new light on those processes within the brain which give rise to such behaviour, and also, by implication, on those patterns of social behaviour that are manifest when numbers of individual minds interact. But there is more than this. By using computers and communication to link many millions of individual human minds in ever more complex patterns, the new research would seem to be moving us daily closer to an emergent human threshold. Just what form this integration will take is up to us, but that it heralds the emergence of some form of global consciousness can hardly still be in doubt.

CHAPTER FIFTY FOUR

10.6 Hypertext

Summary: How we are enabled to deal with 'information overload.'

Hypertext

We've spoken a lot about the power of computers to make information universally available, but already information overload is a well-known problem. Will adding yet more not add to the confusion? How do we ensure that the right information is available to the right people at the right time, without their being swamped by the wealth of data streaming down upon them?

This is a problem with which all living things have had to deal since life began. And it can be solved. Living tissue *selects what it needs* and filters out the rest, responding to just that section of the electro-magnetic spectrum that will help it see, just those wave-lengths that will help it hear, just those magnetic fields that will enable it to navigate - and so on. Selection is as diverse as the information is wide, and every species is different in its needs.

When it comes to the embarrassing avalanche of man-made information threatening to overwhelm mankind, we can do the same. Knowledge - that precious commodity - is piling up so fast we can neither keep it sorted by conventional methods nor make sense of it. Librarians do their heroic best, cataloguing, referencing, cross-referencing. But the limitations of print make the process ever more unwieldy, and the task of finding what you want ever more time-consuming. Thousands of technical journals on thousands of interrelated subjects become ever more difficult to classify, while literally millions of published articles appear each year. And computers compound the problem by spewing forth ever more sheets of specialized data to meet our specialized needs.

Anyone who has ever worked on a council, a commission, a government body or a symposium, will know what mountains of paper-work are entailed. But the weeks and years of delay in normal publication procedures 'slow debate to a crawl.'[1] Moreover, once published, there they are, resistant to modification. Attempts to correct, update or apply them to relative problems are all subject to the same clumsy delays, while parameters change day by day. The moment for decision may long have passed before essential data becomes available. To make matters worse, someone reading the first edition may be unaware that an author has changed his point of view, modified an assertion in the light of subsequent research, or otherwise have chosen to update his work. Nor will the reader know, unless himself a specialist in this author's subject, how his work relates to that of others in the field. Indexes, references, and corrections simply add more pages or more editions, and tracing the links becomes ever more involved.

But while computers help to create the problem, they are also the means to solving it. In the explosion of knowledge that followed the Middle Ages, invention of the printing press ushered in a new age. In the explosion of knowledge in our own century, computers will do the same. They are, in fact, already doing the same. Because of their unique capacity to form associations from amongst astronomical masses of discrete data, computers are ideal instruments for establishing relationships in the network of human knowledge. It is these relationships which form the basis of selectivity in any aspect of social behaviour, whether cultural, political or economic. And it is these relationships which we shall need to clarify in our transition to a self-regulated, non-governmental, non-monetary economy. What, for instance, will be the effect on weather in Bangladesh, when peasants in Argentina burn their rain-forests to bring in cattle for the hamburger market in USA? How much fruit will rot on the quays of Cuba when political disturbance in Eastern Europe disrupts its established market? What is the likely effect of industrial effluent from an industry in Central Europe on the lives of a fishing community on the North Sea? What will fluctuations in the share market in Hong Kong do to the families of migrant workers in Mozambique or Kwazulu/Natal?

Drexler describes the working of a text-handling computer dubbed 'Hypertext' (already on the market), which is specifically programmed for this kind of work.[2] But first he illustrates the concept by the ingenious process of introducing a magician into the 'Chamber of Writings' in which all scholars have thrown their scraps of knowledge in disorderly heaps.

> 'From time to time, some industrous scholar would come to seek knowledge - to rummage through the heap in search of the needed scrap. Some, skilled in such research, could find a particular scrap in as little as a month. The scribes always welcomed researchers: they were so rare.[3]

The researchers have our sympathy. But these imaginary scribes finally get wise. Before sending for the magician, they first turn scholars loose in the chamber with needles and thread to *run strands from scrap to scrap*. The threads have distinguishing colours, as in the treasure-hunts of my childhood, when you only had to follow your own colour to find the treasure. But these scholars are much more specific in what they do. The coloured threads have meanings which are very distinct. Thread of one colour links a scrap *to the next in a series,* another colour might link the same scrap *to a reference,* while yet another might indicate *critical appraisal*. The colours and kinds of linkage are many and various. But, in the end, the scholars have woven a complex web of relationships, represented by a network of coloured threads.

> At last, the magician (with flashing eyes and floating hair) chants a spell, and the whole mess heaves slowly into the air to float like a cloud in the dome. Ever after, a scholar holding a scrap need only tap a thread knotted to its edge to make the linked scrap leap to hand. And the threads, magically, never tangle.
>
> Now the scholars can link scraps on 'The History of the Spread of the Great Plague' to related scraps on history, geography, and medicine. They can add the notes they please, linking them to best advantage. They can add special index scraps, able to bring instantly to hand whatever they list. They can place links where they wish, weaving a network of knowledge to match the connections of the real world.
>
> We, with our inert stacks of paper, could only envy them - if we didn't have computers.[4]

Fortunately, we do have computers, and they are making possible the useful management of information which by its magnitude and complexity might otherwise have overwhelmed us, besides performing a number of tasks for which we were not equipped before.

Today, says Drexler, a number of paper publishers have become electronic publishers, making magazines, newspapers, and journals available through computer networks. And with the right programs, 'text-handling computers will let us link this information in ways even better than magic thread.'[5]

Theodor Nelson, who originated the idea, dubbed the result 'hypertext' to indicate text linked in many directions, not just in a one-dimensional sequence. 'An initial system might be a single-user desktop machine; eventually a growing network of machines will be able to serve as an electronic library.'[6]

> Users will be able to link any part of any document to any other. When a reader points to one end of a link (whether it is shown on the screen by underlining, an asterisk, or a picture of colored thread), the system will fetch and display the material at the other end. Further, it will record new versions of a large document without storing additional copies; it need store only the parts that are changed. This will let it inexpensively store the earlier versions of any document published and modified on the system. It will do all this rapidly, even when the total amount of information stored becomes immense. A network of such machines could eventually mature into a world electronic library.[7]

In a world dependent for its very survival on accurate, interlinked global knowledge, the development of Hypertext must indeed seem like a gift from the Gods. And perhaps the most valuable part of all will be that any idea, continually updated in Hypertext, will be subjected to a cross-fire of references, enabling users to see at a glance whether linked sources support it, or linked criticisms explode it. Drexler points out that this facility will serve to 'skewer nonsense ... clearing it from the intellectual arena'.[8] A valuable tool indeed.

CODA

GAIA
THE BEAUTIFUL
The Cosmic Microbe

* * *

Ours to imagine and to build.
Anything we can imagine, we can create.
Anything we can think, can become reality.
Anything we love, will be worth working for.
Within the framework of her Cosmic Code,
Gaia will build herself anew
upon our dreams
beautiful
living
loving

and Divine

***

Appendix

Extracts from *Chance and Necessity* by Jacques Monod (Collins, 1972).

The way in which allosteric interactions work ... permits a complete freedom in the 'choice' of controls... In short, the very gratuitousness of these systems, giving molecular evolution a practically limitless field for exploration and experiment, enabled it to elaborate the huge network of cybernetic interconnections which makes each organism an autonomous functional unit, whose performances appear to transcend, if not to escape, the laws of chemistry. (p.79)

The analysis of allosteric interactions reveals, first of all, that teleonomic performances are not the exclusive endowment of complex, multi-component systems, since a protein molecule shows itself capable, not only of electively activating a reaction, but of regulating its activity in response to information emanating from several chemical sources.

Secondly, thanks to the concept of gratuity, we are aware how and why these molecular regulating interactions, failing chemical restraints, succeed in being selectively chosen solely for their contribution to the coherence of the system. (p.80)

... the study of these microscopic systems shows us that for complexity, for richness, for potency, the cybernetic network of living beings far surpasses anything that might be predicted by the study of the overall behaviour of whole organisms. And even though these analyses are still far from furnishing us with a complete description of the cybernetic system of the simplest cell, they tell us that, without exception, all activities that contribute to the growth and multiplication of that cell are interconnected and intercontrolled, directly or otherwise. (p.81)

BIBLIOGRAPHY

Allman, N. *Apprentices of Wonder,* Bantam, 1989

Ardrey, R. *The Territorial Imperative*, Collins, 1967

Asimov, I. *Today, Tomorrow*, Abelard-Schuman, 1974

Asimov, I. *A Choice of Catastrophes*, Simon & Schuster, 1979

Autrum, H. 'The Communication Network of the Human Body'and 'The Operational Limits of Animal Sense Organs'in *Man and Animal*, ed. H. Friedrich, MacGibbon and Kie, 1972

Balfour, E. *The Living Soil,* Faber, 1943

Barber, W.J. *A History of Economic Thought*, Penguin, 1970

Barrett, T.J. *Harnessing the Earthworm,* Faber & Faber, 1949

Bateson, G. *Mind and Nature,* Wildwood, 1979

Beckett, D. *Permanent Peace*, Saga, 1985

Bellini, J. *High Tech Holocaust*, David & Charles, 1986

Berg, A. *Malnutrition - What Can Be Done,* John Hopkins University Press, 1987

Bleibtreu, G.N. *The Parable of the Beast: An Exploration of the Frontiers of Biological Knowledge*, Gollancz, 1968

Bodington, S. *Computers and Socialism*, Spokesman, 1973

Bolter, J.D. *Turing's MAN*, Penguin, 1984

Boyle, S. and Ardill, J. *The Greenhouse Effect*, Hodder, 1989

Brand, S. *The Media Lab*, Viking, 1987

Bromfield, L. *Malabar Farm*, Cassell & Company, 1949

Bromley, R. and Gerry, C. (eds) *Casual Work and Poverty in Third World Cities*, John Wiley & Sons, 1979

Brown, L.R. *Building a Sustainable Society*, Norton, 1981

Burton, M. *The Sixth Sense of Animals*, Taplinger Publishing Co., 1973

Calder, N. *The Mind of Man*, BBC, 1970

Capra, F. *The Turning Point*, Flamingo, 1985

Carnoy, M. and Shearer, D. *Economic Democracy*, M.E. Sharpe Inc., 1980

Chisholm, A. ed. *Philosophers of the Earth*, Sidgwick, 1972

Clarke, J. *Our Fragile Land,* MacMillan, 1974

Clements, F. *Bio-Ecology*, Wiley, 1939

Daly, H. *Toward a Steady State Economy*, Freeman, 1973

Davies, P. *Other Worlds*, J.M. Dent & Sons, 1980

de Bono, E. *Conflicts*, Harrap, 1985

de Sola Pool, I. *Technologies of Freedom,* Vintage Books, 1964

Dillard, A. *Pilgrim at Tinker's Creek,* Picador, 1974

Drexler, K.E. *Engines of Creation*, Anchor, 1986

Droscher, V.B. *They Love and Kill,* W.H. Allen, 1977

Drucker, P. *The Frontiers of Management*, Heinemann, 1987

Ehrlich, P. *The Machinery of Nature*, Simon & Schuster, 1986

Ehrlich, P. & A. *Extinction*, Random House, 1981

Ehrlich, P., Sagan, C. & Kennedy, D. *The Nuclear Winter: The World after Nuclear War*, Sidgwick, 1985

Eiseley, L. *The Immense Journey*, Gollancz, 1958
Eiseley, L. *The Firmament of Time*, Gollancz, 1961
Elkington, J. *Sun Traps*, Penguin, 1984
Elkington, J. *The Green Capitalists*, Gollancz, 1987
Ferguson, M. *The Aquarian Conspiracy*, Granada, 1982
Ford, B. *Microbe Power*, Macdonald & James, 1976
Friedrich, H. ed. *Man and Animal,* H. Anthum 'The Communications Network of the Human Body', MacGibbon and Kie, 1972
Fritsch, A.J. *Environmental Ethics*, Anchor Books, 1980
Fromm, E. *Beyond the Chains of Illusion*, Simon & Schuster, 1962
Gardner, M. ed. *The Sacred Beetle: Essays in Science*, OUP, 1985
Goldsmith, E. and Hildyard, N. *Earth, Report 2*, Mitchell Beasley, 1990
Goodfield, J. *An Imagined World,* Hutchinson, 1981
Gourlay, K.A. *Poisoners of the Seas*, Zed, 1988
Griffin, D.R. *Evolutionary Continuity of Mental Experience,* Rockefeller University Press, New York, 1976
Handy, C. *The Age of Unreason*, Arrow Books, 1989
Harrison, P. *The Greening of Africa*, Paladin, 1987
Harrison, P. *The Third World Tomorrow*, Penguin, 1991
Hemming, J. *Individual Morality*, Nelson, 1969
Henderson, H. *The Politics of the Solar Age*, Anchor, 1980
Hirsch, F. *Social Limits to Growth*, Routledge & Kegan Paul, 1977
Howard, A. *An Agricultural Testament,* OUP, 1943
Howard, A. *Farming and Gardening for Health or Disease,* Faber & Faber, 1945
Hoyle, F. and Wickramasinghe, C. *Life Cloud*, J.M.Dent & Sons, 1978
Hoyle, F. and Wickramasinghe, C. *Evolution from Space,* Dent, 1981
Huntley, B.S.R. Sunter, C. *South African Environments into the 21st Century*, Human & Rousseau, 1989
Institute of Economic Affairs *The Unfinished Agenda*, 1986
Jenkins, D. *Job Power: Blue and White Collar Democracy*, Doubleday, 1973
Jonas, D. *Other Senses, Other Worlds*, Cassell, 1976
Jonas, D. and Klein, D. *Manchild: A Study of the Infantilization of Man*, Jonathan Cape, 1971
Jungk, R. and Galtung, J. eds. *Mankind 2000, International Future Research*, 1969
Kobayashi, K. *Computers and Communication*, MIT, 1986
Koehler, O. 'Chance, Necessity and Plan in the Living Universe' and 'Non-verbal thinking' in *Man and Animal,* ed. Friedrich, H., MacGibbon & Kie, 1972
Koestler, A. *Bricks to Babel,* Hutchinson, 1980

Leopold, A.	*A Sand County Almanac*, OUP, 1979
Lopez, B.	*Arctic Dreams*, Macmillan, 1986
Lorenz, K.	*On Aggression*, Methuen & Co. 1966
Lovelock, J.	*Gaia: A New Look at Life on Earth*, OUP, 1979
Luard, N.	*The Last Wilderness*, Hamish Hamilton, 1981
Marais, E.	*The Soul of the White Ant,* Penguin, 1973
Marais, E.	*The Soul of the Ape,* Penguin, 1973
McCarrison, R.	*Nutrition and Health*, Faber & Faber, 1953
Meintjes, S. and Jacques, M.	*The Trial of Chaka Dlamini*, Amagi Books, 1990
Milne, L. and M.	*The Senses of Animals and Men*, Penguin 1962
Monod, J.	*Chance and Necessity*, Collins, 1972
Morris, D.	*Measuring the World's Poor*, Pergamon Press Policy Studies, 1979
Murchie, G.	*The Seven Mysteries of Life*, Rider/Hutchinson, 1979
Ornstein, R. and Thompson, R.F.	*The Amazing Brain*, Chatto & Windus, 1985
Peters, T.	*Thriving on Chaos*, Pan & Macmillan, 1988
Pfeiffer, E.	*The Great American Forest*, Prentice-Hall, 1984
Prigogine, I. and Stengers, T.	*Order Out of Chaos,* Heinemann, 1984
Richard, M.	*The Mystery of Animal Migration*, Constable, 1969
Roberts, A.	*The Self-Managing Environment,* Allison & Busby, 1979
Robertson, J.	*The Sane Alternative*, James Robertson, 1983
Robertson, J.	*Future Work*, Gower/Maurice Temple Smith, 1985
Robertson, J.	*Future Wealth*, Cassell, 1989
Rothman, M.A.	*Cybernetics*, Collins, 1972
Russell, P.	*The Awakening Earth*, Routledge, 1982
Rybczynski, W.	*Taming the Tiger*, Penguin, 1985
Sagan, C.	*The Cosmic Connection*, Hodder & Stoughton, 1973
Sagan, C.	*The Cosmic Connection*, Hodder & Stoughton, 1974
Sagan, C.	*The Dragons of Eden*, Hodder & Stoughton, 1977
Sale, K.	*Dwellers in the Land,* Sierra Club Books, 1985
Schumacher, E.F.	*Small is Beautiful,* Blond & Briggs, 1973
Servan-Schreiber, J.J.	*The World Challenge*, Simon & Schuster, 1981
Seymour, J. and Girardet, H.	*Far From Paradise,* BBC, 1986
	Blueprint for a Green Planet, Dorling Kindersley, 1987
Sheldrake, R.	*A New Science of Life: The Hypothesis of Formative Causation,* Hodder, 1973
Sieghart P. ed.	*Microchips with Everything*, Comedia, 1982
Smith, A.	*The Geopolitics.of Information*, Faber & Faber, 1980
Speiser, S.	*A Piece of the Action*, Van Nostrand Reinhold, 1977
Sykes, F.	*Humus and the Farmer*, Faber & Faber, 1951
Teilhard de Chardin, P.	*The Phenomenon of Man*, Collins, 1959
Thomas, L.	*The Lives of a Cell,* Bantam, 1974
Thomas, L.	*The Medusa and the Snail*, Viking Press, 1979
Thomas, L.	*Late Night Thoughts*, OUP, 1985

Thruelsen, R. and Kohler, J. eds. *Adventures of the Mind*, Victor Gollancz, 1960

TIME *Environmental Conference*, November 1988

Tinbergen, J. Co-ord. *Reshaping the International Order: A Report to the Club of Rome*, Hutchinson, 1976

Tinbergen, N. *Social Behaviour in Animals*, Methuen, 1965

Toffler, A. *Future Shock*, Bodley Head, 1970

Toffler, A. *The Third Wave*, Pan, 1981

Toffler, A. *Previews and Premises,* Pan, 1984

Toffler, A. *Power Shift*, Bantam Books, 1990

Tompkins, P. and Bird, C. *The Secret Life of Plants*, Penguin, 1974

Turner, R. *The Eye of the Needle*, Orbis Books, 1978

Van der Wee, H. *Prosperity & Upheaval: The World Economy*, 1945-1980, Viking, 1986

Van Lawick-Goodall, J. *In the Shadow of Man*, Collins, 1971

Waddington, C.H. *Tools for Thought*, Jonathan Cape, 1977

Walter, W.G. *The Living Brain*, Penguin, 1961

Watson, L. *Supernature*, Hodder, 1974

Watson, L. *Lifetide*, Hodder, 1974

Watson, L. *Gifts of Unknown Things*, Hodder, 1976

Watson, L. *The Romeo Error*, Coronet, 1976

Watson, L. *Beyond Supernature*, Hodder, 1986

Weber, J. *Tomorrow's World,* Computers, BBC, 1984

Weber, R. *Dialogues with Scientists and Sages,* Routledge & Kegan Paul, 1986

Weinberg, S. *The First Three Minutes*, Andre Deutsch, 1977

Wenk, E. Jr. *Margins for Survival,* Pergamon Press, 1979

White, J. *The Politics of Foreign Aid,* Bodley Head, 1974

Williams, R. *Towards 2000*, Penguin, 1985

Wilsher, P. and Righter, R. *The Exploding Cities,* Deutsch, 1975

Notes

PART 1

Chapter 1

1 Alfred Lord Tennyson, *The High Pantheism VI.*

PART 2

Chapter 2

1 Mach's Principle as quoted by Arthur Koestler, *Bricks to Babel,* Ch. 2.
2 David Bohm, British Physicist. Quoted by Peter Russell, *The Awakening Earth.*
3 *Ibid.*
4 Peter Russell, *Ibid*, Ch.8 p.126.
5 Lyall Watson, *Gifts of Unknown Things.*
6 *Ibid.*
7 Bell's Theorem. Quoted by Arthur Koestler in *Ibid,* Ch.52 p.660.

Chapter 3

1 Lyall Watson, *Supernature.*
2 James E. Lovelock, *Gaia. A New Look at Life on Earth.*
3 *Ibid.*
4 *Ibid.*
5 Lyall Watson, *Gifts of Unknown Things,* p.33.

Chapter 4

1 Peter Russell, *The Awakening Earth.*
2 Ilya Prigogine & Isabelle Stengers, *Order Out of Chaos.*
3 *Ibid.*
4 Fritjof Capra, *The Turning Point.*
5 Ilya Prigogine & Isabelle Stengers, *Ibid.*
6 Fritjof Capra, *Ibid.*
7 Ilya Prigogine and Isabelle Stengers, *Ibid.*
8 Peter Russell, *Ibid.*
9 *Ibid.*
10 *Ibid.*
11 Ilya Prigogine and Isabelle Stengers, *Ibid.*
12 *Ibid.*

PART 3

Chapter 5

1 Doris Jonas and David Jonas, *Other Senses, Other Worlds.*

Chapter 6

1 Brian Ford, *Microbe Power.*
2 *Ibid.*
3 Carl Sagan, *The Dragons of Eden.*
4 Brian Ford, *Microbe Power.*
5 *Ibid.*
6 *Ibid.*
7 *Ibid.*
8 Ilya Prigogine & Isabelle Stengers, *Ibid.*
9 Brian Ford, *Ibid.*
10 *Ibid.*

Chapter 7

1 Brian Ford, *Microbe Power.*
2 Carl Sagan, *The Dragons of Eden.*
3 *Ibid.*

PART 4

Chapter 8

1 Fred Hoyle and Chandra Wickramasinghe, *Evolution from Space.*
2 Ilya Prigogine and Isabelle Stengers, *Order Out of Chaos.*
3 Stephen Hawking, *Beyond the Black Hole. (Stephen Hawking's Universe,* by John Boslough).
4 Ilya Prigogine and Isabelle Stengers, *Ibid.*

Chapter 9

1 Jacques Monod, *Chance and Necessity,* p.80.
2 *Ibid.*
3 *Ibid.*
4 Ilya Prigogine and Isabelle Stengers, *Order Out of Chaos.*
5 Jacques Monod, *Chance and Necessity.*

Chapter 10

1 Ilya Prigogine and Isabelle Stengers, *Order Out of Chaos.*
2 Lewis Thomas, *The Lives of a Cell.*
3 *Ibid.*
4 *Ibid.*
5 *Ibid.*
6 *Ibid.*
7 Jacques Monod, *Chance and Necessity.*
8 J.E. Lovelock, *Gaia. A New Look at Life on Earth*, p.57.

Chapter 11

1 Rupert Sheldrake, *A New Science of Life. The Hypothesis of Formative Causation.*
2 Lyall Watson, *Supernature.*
3 Rupert Sheldrake, *Ibid.*
4 *Ibid.*
5 *Ibid.*
6 *Ibid.*

Chapter 12

1 Carl Gustav Jung, as quoted by Sheldrake, *Ibid.*
2 Rupert Sheldrake, *A New Science of Life. The Hypothesis of Formative Causation.*
3 Rupert Sheldrake, *Ibid.*
4 Lyall Watson, *Lifetide,* p.167.
5 *Ibid.*
6 *Ibid,* p.156.
7 Carl Gustav Jung, as quoted by Sheldrake, *Ibid.*

Chapter 13

1 Fred Hoyle and Chandra Wickramasinghe, *Evolution from Space.*
2 S. Ohno, *Evolution by Gene Duplication* as quoted by Fred Hoyle and Chandra Wickramasinghe, *Ibid.*
3 Peter Russell, *The Awakening Earth.*

PART 5

Chapter 14

1 David Bohm, British Physicist. As quoted by Peter Russell in *The Awakening Earth.*
2 Bell's Theorem. Quoted by A. Koestler in *Bricks to Babel,* Ch. 52. p.660.
3 Jacques Monod, *Chance and Necessity,* p.80.
4 Jack Weber, *Computers,* p.106
5 Peter Russell, *The Awakening Earth.*
6 Fritjof Capra, *The Turning Point*, p.26.
7 *Ibid.*

Chapter 15

1 Robert Ornstein and Richard F. Thompson, *The Amazing Brain.*
2 *Ibid.*

Chapter 16

1 Richard E. Farson. Monograph: Bill of Rights for 1984. Taken from *Mankind 2000.* Editors Robert Jungk and Johan Galtung (International Future Research, London, 1969).
2 Hansjochem Autrum. Essay on 'The Communications Network of the Human Body' in *Man and Animal,* Edited by Heinz Friedrich..
3 *Ibid.*
4 *Ibid.*
5 Rutherford Platt, *The Great American Forest.*
6 *Ibid.*
7 *Ibid.*
8 *Ibid.*

Chapter 17

1 Otto Koehler, *Essay:* 'Chance, Necessity and Plan in the Living Universe' in *Man and Animal,* Edited by Heinz Friedrich.
2 *Ibid.*
3 *Ibid.*
4 *Ibid.*
5 *Ibid.*

PART 6

Chapter 19

1 *Mankind 2000,* Editors Robert Jungk and Johan Galtung.
2 *Ibid.*
3 *TIME Environmental Conference*, Nov. 1988. Thomas A. Sancton. Paper on Multinational Co-operation. (Published in TIME Magazine 2 January 1989.)
4 *Ibid.*
5 *TIME Environmental Conference,* Nov. 1988 Paper by Michael Lemonick, quoting Stephen Schneider of the Nat. Centre for Atmospheric Research.
6 Thomas Livejoy of the Smithsonian Institution, *TIME Environmental Conference*, Nov. 1988.
7 Thomas A. Sancton, *Ibid.*
8 Robert Jungk, Preface, *Mankind 2000, ibid.*

Chapter 21

1 Frederic E. Clements, of the Carnegie Institute in: *Bio-Ecology.*
2 *Ibid.*
3 *Ibid.*
4 Paul and Ann Ehrlich, *Extinction.*
5 *The Nuclear Winter: The World After Nuclear War*, Paul R. Ehrlich, C. Sagan, Donald Kennedy, Walter O. Roberts.
6 Thomas A. Sancton, Essay: 'Facing up to Impending Disaster.' *TIME Environmental Conference,* 1988, *ibid.*
7 *Ibid.*
8 *Ibid.*

Chapter 22

1 K. Eric Drexler, *Engines of Creation*, Ch.7 'Engines of Healing'.
2 Michael Shallis, *The Silicon Idol*, p.169.
3 *Ibid*, p.177.
4 Witold Rybczynski, *Taming the Tiger.*
5 Arthur Schopenhauer as quoted by Lyall Watson in *Beyond Supernature*.
6 Alvin Toffler, *Future Shock*.
7 Alvin Toffler, *The Third Wave*.
8 Alvin Toffler, *Previews and Premises.*
9 E.F. Schumacher, *Small is Beautiful.*
10 Jean-Jacques Servan-Schreiber, *The World Challenge.*
11 *Ibid*.
12 *Ibid.*

Chapter 23

1 E.F. Schumacher, *Small is Beautiful.*
2 Peter Wilsher and Rosemary Righter, *The Exploding Cities.*
3 James Hemming, *Individual Morality.*
4 Peter Wilsher and Rosemary Righter, *Ibid.*
5 James Hemming, *Ibid.*

Chapter 24

1 Raymond Williams, *Towards 2000.* p.268.
2 Jan Tinbergen, (Co-ordinator) Reshaping the International Order. Rio. A Report to the Club of Rome.
4 *Shorter Oxford Dictionary.*
5 Lewis Thomas, *Late Night Thoughts*.
6 Adapted from Radio S.A. *Special Report,* 13 November 1989. Speaker: Senior Council Ishmail Mahommed, Judge of Appeal in Lesotho & Swaziland.
7 *The Citizen,* South African Daily Newspaper.

Chapter 25

1 Herman Daly, *Toward a Steady State Economy.*
2 Hazel Henderson, *The Politics of the Solar Age.*
3 Staffan Linder, as quoted by Daly in *ibid.*

Chapter 26

1 Fritjof Capra, *The Turning Point. Science, Society and the Rising Culture.*

PART 7

Chapter 28

1 William J. Barber, *A History of Economic Thought*.
2 *Ibid*.
3 *Ibid*.
4 Adam Smith, *An Inquiry into the Nature and Causes of the Wealth of Nations*, 1776.
5 *Ibid,* p.28.
6 *Ibid,* p.30.
7 Hazel Henderson, *The Politics of the Solar Age.*
8 F.A. Hayek. Essay: 'The Moral Imperative', *The Unfinished Agenda*.
9 *Ibid.*
10 Stephen Bodington, *Computers and Socialism.*
11 William J. Barker, *A History of Economic Thought,* p.30.
12 *Ibid*, p.30.
13 *Ibid*, p.31.
14 *Ibid*, p.84.
15 *Ibid*, p.97.
16 *Ibid*, p.97.
17 *Ibid*, p.131.
18 *Ibid.*
19 Michael Shallis, *The Silicon Idol.*
20 *Ibid*, p.58 Quoting Joseph Weizenbaum, *Technological Intoxication* p.305.

Chapter 29

1 F.A. Hayek. Essay: 'The Moral Imperative', *The Unfinished Agenda.*
2 *Ibid.*

Chapter 31

1 Jack Weber, *Computers.*
2 *Ibid*.
3 *Ibid*.
4 *Ibid*.

Chapter 32

1 Alan Smith, *The Planned Economies of Eastern Europe*.
2 *Ibid*, p.85.

Chapter 33

1 Jack Weber, *Computers*.

PART 8

Chapter 35

1 Stephen Bodington, *Computers and Socialism.*
2 *Ibid*.
3 *Ibid*.
4 *Ibid*.
5 Stephen Bodington, *Ibid.*
6 *Ibid*.
7 *Ibid*.
8 *Ibid*.

9 *Ibid.*

Chapter 36

1 Jean-Jacques Servan-Schreiber, *The World Challenge*.
2 Rutherford Platt, *The Great American Forest.*
3 *Ibid.*

BOOK III

PART 9

Chapter 39

1 P. Drucker, *The Frontiers of Management* p.261.
2 *Ibid* p.261.
3 *Ibid* p.21.
4 T. Peters, *Thriving on Chaos.*
5 *Ibid* p.xi.
6 *Ibid* (From Preface).
7 S. Brand, *The Media Lab* p.11.
8 *Ibid* p.46.
9 G. Bateson, *Mind and Nature*, p.66.
10 S. Boyle, *The Greenhouse Effect* (quoting p.93 Ardill J. Shridath Ramphal, Feb.1989).

Chapter 40

1 J. Elkington,*The Green Capitalists*.
2 K.A. Gourlay, *Poisoners of the Seas* p.238.
3 *Ibid,* p.16.
4 *Ibid,* p.14.
5 *Ibid,* p.16.
6 *Ibid,* p.16.
7 *Ibid,* p.16.
8 F. Kinsman, *Resurgence*. Diary. No.133.
9 J. Elkington, *Ibid*, p.125.
10 *Ibid* p.126.
11 R. Dubos, *Philosophers of the Earth,* p.15. Conversation with Ann Chisholm.
12 D. Kuenen, *Philosophers of the Earth,* p.94, Conversation with Ann Chisholm.
13 *Ibid*, p.96.
14 J. Seymour & H. Girardet, *Blueprint for a Green Planet*.
15 *Ibid*, p.27.
16 *Ibid*, p.29.
17 *Ibid*, p.30.
18 *Ibid*, p.30.
19 E. Balfour, *The Living Soil,* p.14.
20 S. Boyle, *The Greenhouse Effect*, p.92.
21 A. Howard, *Testament of Agriculture,* p.2.
22 *Ibid,* p.2.
23 *Ibid,* p.136.
24 *Ibid*, p.3.
25 *Ibid,* p.19.
26 E. Balfour, *Ibid*, p.47.
27 *Ibid*, p.48.
28 A. Howard, *Ibid*, p.161.
29 E. Balfour, *Ibid,* p.52.
30 *Ibid,* p.52.
31 *Ibid* (Quoting: King, *Farmers of Forty Centuries*, p 52).
32 P. Drucker, *The Frontiers of Management,* p.x Preface.

Chapter 41

1 S. Meintjes, *The Trial of Chaka Dlamini* p. 5 & M. Jacques, *An Economic Scenario for the New South Africa.*
2 *Ibid,* p.5.
3 *Ibid*, p.9.
4 *Ibid,* p.11.
5 *Ibid,* p.13.
6 *Ibid,* p.14.
7 *Ibid,* p.15.
8 *Ibid,* p.61.

Chapter 43

1 D. Beckett, *Permanent Peace,* p.11.
2 *Ibid,* p.96.
3 *Ibid,* p.12.
4 *Ibid,* p.12.
5 *Ibid,* p.13.
6 *Ibid,* p.13.
7 *Ibid,* p.14.
8 *Ibid,* p.145.
9 *Ibid*, p.136.
10 *Ibid,* p.75.
11 R. Heller, *Culture Shock,. p.59.*
12 *Ibid,* p.59
13 P. Wilsher & R. Righter, *The Exploding Cities,* p.127.

Chapter 44

1 I. Prigogine, *Order Out of Chaos,* p.189.

Chapter 45

1 E. Goldsmith, *Earth Report 2.*
2 *Ibid,* p.158.

Chapter 46

1 P. Drucker, *The Frontiers of Management*, p.265.
2 A. Tennyson, *Mort d'Arthur*.
3 Operation Hunger - 'News'
4 D. Jenkins, *Job Power,* p.248.
5 *Ibid*, p.249.
6 *Ibid*, p.256.
7 E. Schumacker, *Small Is Beautiful.* p.62.
8 *Ibid*, p.62.

Chapter 47

1 T. Peters, *Thriving on Chaos,* p.25.
2 *Ibid,* p.31.
3 *Ibid,* p.31.
4 P. Drucker, *The Frontiers of Management,* p.173.
5 *Ibid,* p.173.
6 S. Brand, *The Media Lab,* p.24.
7 T. Peters, *Ibid,* p.302.
8 *Ibid,* p.302.

Chapter 48

1 J. Elkington, *Sun Traps,* p.23.
2 *Ibid*, p.23
3 *Ibid,* p.171.
4 *Ibid,* p.171.
5 *Ibid,* p.173
6 *Ibid,* p.173.
7 E. Goldsmith, *Earth Report 2,* p.63 & N. Hildyard.
8 K. Gourlay, *Poisoners of the Seas,* p.240.
9 J. Elkington, *Sun Traps,* p.24.
10 *Ibid* (Quoting A. Lovins), p.24.
11 E. Goldsmith, *Earth Report 2,* p.64 N. Hildyard.
12 *Ibid*, p.64.
13 *Ibid*, p.64.
14 P. Drucker, *The Frontiers of Management,* p.27.
15 R. Platt, *The Great American Forest,* p.113.
16 *Ibid*, p.113.
17 K. Drexler, *Engines of Creation,* p.94.
18 J. Elkington, *Sun Trap*, p.33.
19 R. Platt. *The Great American Forest,* p.114.
20 *Ibid,* p.39.
22 J. Elkington, *Sun Traps,* p.159
23 *Ibid* p.159.
24 *Ibid* p.160.
25 *Ibid* p.160.
26 J. Elkington, *Sun Traps,* p.34
27 B. Ford B, *Microbe Power,* p.17.
28 *Ibid,* p.39.
29 J. Elkington, *Ibid,* p.48.
30 K. Gourlay, *Poisoners of the Seas,* p.72.
31 *Ibid,* p.70.
32 *Ibid,* p.73.
33 E. Goldsmith, *Earth Report 2.*
34 *Ibid,* p.74.

PART 10

Chapter 49

1 K. Drexler, *Engines of Creation*, p.96.
2 *Ibid,* p.233.
3 *Ibid,* p.201.

Chapter 50

1 K. Drexler, *Engines of Creation,* p.236.
2 K. Kobayashi, *Computers & Communication*, p.xi.

Chapter 52

1 K. Kobayashi, *Computers & Communication,* C&C, p.110.

Chapter 53

1 K. Drexler, *Engines of Creation,* p.77.
2 W. Allman, *Apprentices of Wonder,* p.12.
3 *Ibid,* p.82.
4 *Ibid,* (Quoting J. Hopfield) p.11.
5 *Ibid,* p.79.

Chapter 54

1 K. Drexler, *Ibid,* p.218.
2 *Ibid,* p.220.
3 *Ibid,* p.218.
4 *Ibid*, p.219.
5 *Ibid,* p.220.
6 *Ibid,* p.220.
7 *Ibid,* p.221.
8 *Ibid,* p.224.

Index